NMR of
Aromatic Compounds

NMR of Aromatic Compounds

J. D. MEMORY

Dean, the Graduate School
Vice Provost
Professor of Physics
North Carolina State University
Raleigh, North Carolina

NANCY K. WILSON

U.S. Environmental Protection Agency
Research Triangle Park, North Carolina

A Wiley-Interscience Publication

JOHN WILEY & SONS

New York • Chichester • Brisbane • Toronto • Singapore

Library of Congress Cataloging in Publication Data:

Memory, Jasper D.
  NMR of aromatic compounds.

  Includes bibliographic references and index.
    1. Nuclear magnetic resonance spectroscopy.   2. Aro-
matic compounds—Spectra.   I. Wilson, Nancy K. (Nancy
Keeler), 1937– . II. Title.   III. Title: N.M.R. of
aromatic compounds.
QD331.M45   1982   547′.6046   82-8588
ISBN 0-471-08899-4   AACR2

Printed in the United States of America

10 9 8 7 6 5 4 3 2 1

Preface

This book is designed for organic chemists at the graduate and professional levels, and provides a survey of the application of high-resolution NMR to the study of aromatic compounds. The field is vast, and it has been necessary to make certain selections regarding the topics covered. It is hoped, however, that the material has been presented in a useful format, and that sufficient references that will help the reader obtain further information have been provided.

Most high-resolution NMR on aromatic compounds has been done with protons or ^{13}C nuclei, and these studies receive most of the attention here, but other nuclei are also discussed. Theories of the NMR parameters (chemical shifts, coupling constants, and relaxation times) have been treated in sufficient detail to enable the reader to make sense of most of the current literature. Experimental techniques in the field are discussed in approximately the same depth, and many experimental data are included as well.

Various books on NMR in general and on certain specific areas of NMR have appeared over the years, but this volume is, to our knowledge, the only book that concentrates exclusively on aromatic compounds. Considerable attention has been given to polycyclic aromatic compounds as prototypes of the compounds dealt with. Substituent effects are dealt with in some detail, so that the results stated here should have wide applicability.

J. D. MEMORY
NANCY K. WILSON

Raleigh, North Carolina
Research Triangle Park, North Carolina
September 1982

v

Acknowledgments

Permission is gratefully acknowledged for the use of copyrighted material as follows:

Figures 1.1 and 1.2 from *Quantum Theory of Magnetic Resonance Parameters*, published by McGraw-Hill Book Company.

Figures 2.2, 2.3, 4.2, 4.4, 4.7, 5.1, 7.8, and 7.9 from the *Journal of Magnetic Resonance*, published by Academic Press.

Figures 2.1, 2.4, and 4.1 from *High Resolution Nuclear Magnetic Resonance*, published by McGraw-Hill Book Company.

Figures 3.1 and 3.2 from *Molecular Physics*, published by Taylor and Francis, Ltd.

Figures 3.3, 3.4, and 3.6 from the *Journal of Chemical Physics*.

Figure 3.5 from the *Journal of Molecular Spectroscopy*, published by Academic Press.

Figures 4.3, 6.7, 6.8, and Table 5.11 from the *Journal of the American Chemical Society*, published by the American Chemical Society.

Figures 4.5 and 4.9 from the *Canadian Journal of Chemistry*, published by the National Research Council of Canada.

Figure 4.6 from *Mass Spectrometry and NMR Spectroscopy in Pesticide Chemistry*, published by Plenum Press.

Figures 4.8 and 6.6 from the *Journal of Organic Chemistry*, published by the American Chemical Society.

Figure 6.9 from *Helvetica Chemica Acta*, published by the Swiss Chemical Society.

Figure 6.10 from the *Bulletin of the Chemical Society of Japan*, published by the Chemical Society of Japan.

Figure 6.11 from *Tetrahedron Letters*, published by Pergamon Press.

Table 7.2 from the *Journal of the Chemical Society, Perkin Transactions II*, published by the Royal Society of Chemistry.

Additionally, we thank R. Freeman, H. Günther, R. Mondelli, L. Lunazzi, D. H. Williams, J. A. Pople, J. F. M. Oth, O. Yamamoto, H. Nakanishi, M. J. S. Dewar, R. B. Mallion, J. M. Gaidis, H. J. Jakobsen, and K. Hayamizu for permission to reproduce parts of their work.

This work was not funded by the U.S. Environmental Protection Agency. The contents do not necessarily reflect the views of the Agency and no official endorsement should be inferred.

J.D.M.
N.K.W.

Contents

NMR of
Aromatic Compounds

One

Fundamental Aspects of NMR

THE RESONANCE CONDITION

Nuclei with nonzero spin have a magnetic moment, $\boldsymbol{\mu}$, related to the spin by

$$\boldsymbol{\mu} = \hbar\gamma\mathbf{I} \tag{1.1}$$

where $\hbar$ is Planck's constant divided by 2π, $\mathbf{I}$ is the spin in units of $\hbar$, and γ is the magnetogyric ratio of the nucleus, defined by Eq. 1.1. In a magnetic field $\mathbf{H}_0$, assumed to be in the z direction, the energy of the nuclear magnetic moment is given by

$$E = -\boldsymbol{\mu} \cdot \mathbf{H}_0 = -\hbar\gamma H_0 I_z \tag{1.2}$$

For a nucleus of spin I, the eigenvalues of I_z are $I, I - 1, \ldots, -I$; that is, I_z can take on values from I to $-I$ in steps of one. Protons, ^{19}F nuclei, and ^{13}C nuclei have $I = 1/2$, for example; deuterons and ^{14}N nuclei have $I = 1$, and so forth.

Electromagnetic radiation of the proper polarization and with frequency ν satisfying the Bohr rule

$$h\nu = \Delta E \tag{1.3}$$

can induce transitions between the energy levels of the system; the process is called *nuclear magnetic resonance* (NMR).

Combining Eqs. 1.2 and 1.3 we obtain, for the resonance absorption frequency,

$$\nu = \gamma H_0 \Delta I_z / 2\pi \tag{1.4}$$

The selection rule

$$|\Delta I_z| = 1$$

yields

$$\nu = \left(\frac{1}{2\pi}\right) \gamma H_0 \tag{1.5}$$

for the resonance condition.

In the simplest experimental arrangement to observe NMR, the sample containing the nuclei to be investigated sits between the poles of a large magnet, which provides the external magnetic field required for the experiment (see Figure 1.1), and within several turns of wire that form a helix with its axis perpendicular to the external magnetic field. The coil itself is the inductive part of a tuned LC circuit of a radio frequency (rf) oscillator. When the oscillator is operating, there is an alternating magnetic field, small in magnitude compared to the external field, along the axis of the coil. If the frequency of this oscillation coincides with the resonant frequency of the nuclei—that is, if the frequency satisfies the basic resonance condition given as Eq. 1.5—NMR occurs. Since NMR involves a realignment of some of the nuclear spins to a direction opposite to the magnetic field, corresponding to a higher energy state for the spin system, there will be an absorption of energy by the nuclear spin system from the source of the alternating magnetic field. This absorption of energy can be detected electronically and then displayed on either an oscilloscope or a graphic recorder. One may either keep the frequency of the oscillator constant and vary the magnetic field strength until Eq. 1.5 is satisfied or keep the field strength constant and vary the frequency.

A different experimental arrangement for observing NMR involves the use of a second coil at right angles to both the external magnetic field and the first coil that provides the rf. Classically, it may be shown that the effect of the rf at resonance is to tip the nuclear magnetization vector away from alignment with the external magnetic field. This magnetization will then precess with the re-

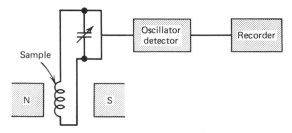

Figure 1.1. Block diagram of a simple NMR spectrometer.

sonance frequency given by Eq. 1.5. This precessing nuclear magnetization will induce a signal at the resonance frequency in the second coil, and the signal may be detected, amplified, and displayed. A spectrometer operating on this principle is called a nuclear induction spectrometer.

Both types of experiment just described are continuous wave (CW) experiments; that is, the rf oscillator providing the alternating magnetic field operates continuously. There are NMR spectrometers that are pulsed: the resonant radiation is in the form of a brief, intense pulse, before and after which there is no rf provided by the oscillator. To see why pulsed NMR spectroscopy is useful, one must consider the problem of the signal-to-noise (S-N) ratio for CW spectrometers.

The sensitivity of a spectrometer is directly related to the S-N ratio. The standard technique for improving the S-N ratio is to obtain many spectra of the same sample and then add these spectra by computer, a process known as time-averaging computer improvement of the S-N ratio. The NMR signal will build up coherently as many spectra as are added, whereas the noise, which is random, will tend to average out. To obtain a typical CW spectrum requires rather a long time, of the order of minutes. One reason is that most spectrometers incorporate an electronic integrating circuit to cut down noise, and the time constant of that circuit provides a lower bound on the time required to record the spectrum. Obtaining many spectra, then, each of which requires a good deal of time, can make time-averaging computer improvement of the S-N ratio quite a tedious and time-consuming process. Another reason that typical CW spectrometers are slow is that one must scan regions of the spectrum that have no resonances.

The basic principle of a pulsed spectrometer is that the nuclear induction signal following an intense rf pulse at the resonance frequency of the nuclei contains all the information necessary to reproduce the NMR spectrum of the system. Indeed, it can be shown that the nuclear induction signal arising from the precessing nuclei following a pulse at the resonant frequency, called a free induction decay (FID), is simply the Fourier transform of the CW spectrum. Spectrometers operating on this principle are called FT-NMR spectrometers. The great advantage of FT-NMR spectroscopy is that the time required to obtain a spectrum is much less than that required to obtain a CW spectrum. The limiting factor for an FT spectrometer is the relaxation time of the nuclear spin system (see the section on time-dependent phenomena in this chapter), not the time constant of an electronic integrating circuit in the spectrometer. For this reason, a large number of spectra can be obtained in a relatively short time, those spectra can be added, and high sensitivities can be achieved. It is the advent of FT-NMR spectroscopy that has made natural abundance ^{13}C NMR spectroscopy a practical tool for the analytical chemist. Even the feeble signal from ^{13}C nuclei in natural abundance in a sample can be enhanced dramatically by obtaining many spectra quickly and adding them.

HIGH-RESOLUTION NUCLEAR MAGNETIC
RESONANCE SPECTRA

The resonant frequency of a magnetic nucleus is proportional to the magnetic field in which the nucleus finds itself. Since nuclei have magnetic moments, they themselves produce a magnetic field. In a typical solid, the local magnetic field at the site of a nucleus includes these magnetic fields produced by neighboring nuclei and thus can vary over a range of several gauss. In nonviscous liquids, however, the molecules move rapidly and randomly. This motion has the effect of averaging to zero the line broadening caused by the dipole–dipole interaction among nuclear magnetic moments. This makes possible quite sharp lines, roughly one Hz in width, so that spectra under these conditions are called *high-resolution* NMR spectra. High resolution leads to the observation of spectral structure, typically because of two phenomena: *chemical shifts* or *shielding*, and *spin–spin coupling*. The first occurs when the local magnetic field arising from electronic currents induced by the magnetic field, H_0, differs from one nuclear site to another in a molecule, so that the resonance frequencies, according to Eq. 1.5, differ slightly. There is then a "chemical shift" between the absorption lines arising from different nuclei.

Since the electronic currents, and hence the secondary local magnetic field accounting for shielding, are proportional to the strength of the external magnetic field, H_0, one can write

$$H' = -\sigma H_0 \qquad (1.6)$$

for the strength of the local field, H'; the constant of proportionality, σ, is called the *shielding parameter* and the minus sign in Eq. 1.6 comes from the fact that the shielding is usually diamagnetic; that is, the secondary induced field is opposite to the external field.

The resonance frequency of a nucleus with shielding parameter σ in an external field, H_0, is, by generalization of Eq. 1.5,

$$\nu = (2\pi)^{-1}\gamma(H_0 + H') = (2\pi)^{-1}\gamma H_0(1 - \sigma) \qquad (1.7)$$

It is clear that line separations in high-resolution NMR spectra that are caused by chemical shifts—that is, to differences in shielding—will be proportional to the strength of the magnetic field H_0 in the experiment.

Spectral structure caused by spin–spin coupling comes from an indirect scalar interaction between two nuclear magnetic moments through mutual interaction with the magnetic moments of electrons; since this interaction has a scalar form, it does not average to zero through molecular reorientation as does the direct dipole–dipole interaction, which has a tensor form. It is shown in Chapter 5 that the form of spin–spin interaction is

$$V_{ss} = hJ_{12}\mathbf{I}_1 \cdot \mathbf{I}_2 \qquad (1.8)$$

where J_{12} is the spin–spin coupling constant between nuclei 1 and 2. The splitting of lines caused by this interaction does not depend on H_0 as does spectral structure caused by chemical shifts. This difference can be helpful in analyzing an NMR spectrum.

For a molecule with N nuclei, with magnetogyric ratios γ_i, shielding parameters σ_i, and coupling with one another with spin–spin coupling constants $J_{ii'}$, the spin Hamiltonian in a magnetic field H_0 in the z direction is given by

$$\mathcal{H} = \frac{-hH_0}{2\pi} \sum_i \gamma_i(1 - \sigma_i) I_{iz} + h \sum_{i<i'} J_{ii'} \mathbf{I}_i \cdot \mathbf{I}_{i'} \tag{1.9}$$

where we have used Eqs. 1.2, 1.7, and 1.8.

The energy levels of the spin system are determined by finding the eigenvalues of $\mathcal{H}$, and the line positions and intensities are determined by the eigenvalues and their corresponding eigenvectors. We now outline this procedure.

Spin functions consisting of products of single-particle eignvectors of I^2 and I_z are used as a starting basis set. For a set of two spin 1/2 nuclei, there would be four such: $\alpha(1)\alpha(2)$, $\alpha(1)\beta(2)$, $\beta(1)\alpha(2)$, and $\beta(1)\beta(2)$, where α and β refer to spin "up" ($I_z = 1/2$) and spin "down" ($I_z = -1/2$) states of a single spin 1/2 particle, where "up" corresponds to a spin aligned with H_0 and "down" corresponds to a spin aligned opposite to H_0.

With respect to such a basis, a matrix representation of $\mathcal{H}$ can be written and diagonalized. The eigenvalues, E_i, and eigenvectors, V_i (which will be linear combinations of the basis vectors), are determined in the diagonalization process. The frequency of absorbed NMR radiation corresponding to a transition from state V_j to V_i is, of course, just

$$\nu_{ij} = \frac{E_i - E_j}{h} \tag{1.10}$$

Some transitions, however, are forbidden; one may also wish to predict the intensity of a particular absorption line. Time-dependent perturbation theory, used in conjunction with knowledge of the eigenvectors, V_i, sheds light on both these matters. The perturbation leading to NMR transitions is the coupling of the nuclear magnetic moments with the oscillating weak magnetic field at right angles, along the x axis for example, to the large constant magnetic field H_0, which is taken to be in the z direction. It can be shown that this perturbation leads to an expression for the intensity of a transition between states i and j that has the form

$$P_{ij} \sim \left| \left\langle i \left| \sum_k^N \gamma I_{kx} \right| j \right\rangle \right|^2 \tag{1.11}$$

where I_{kx} is the x component of the spin of the kth nucleus, and the matrix element is taken between states specified by eigenvectors V_i and V_j.

It can be shown that Eq. 1.11 leads directly to a selection rule:

$$\Delta\left(\left\langle i\,\middle|\,\sum_{k=1}^{N} I_{kz}\,\middle|\,j\right\rangle\right) = \pm 1 \tag{1.12}$$

That is, there are no NMR transitions except between states differing by exactly one unit in total z component of spin.

The relative line intensities allowed can be determined by evaluating the quantity on the right-hand side of Eq. 1.11, which will depend on $\mathbf{V}_i$ and $\mathbf{V}_j$.

As an example, consider the case of a system of two spin 1/2 nuclei with magnetogyric ratio γ, with shielding parameters σ_1 and σ_2, and coupled with a spin–spin interaction of coupling constant J. One first defines the chemical shift between the nuclei by

$$\delta = \sigma_1 - \sigma_2 \tag{1.13}$$

In an external magnetic field, $\mathbf{H}_0$, the Hamiltonian of the system is, from Eq. 1.9,

$$\mathcal{H} = \frac{-h}{2\pi}\,\gamma H_0[(1 - \sigma_1 I_{1z}) + (1 - \sigma_2 I_{2z})] + hJ\mathbf{I}_1 \cdot \mathbf{I}_2 \tag{1.14}$$

The chosen basis set is

$$\mathbf{V}_1 = \alpha(1)\alpha(2)$$
$$\mathbf{V}_2 = \alpha(1)\beta(2)$$
$$\mathbf{V}_3 = \beta(1)\alpha(2)$$
$$\mathbf{V}_4 = \beta(1)\beta(2) \tag{1.15}$$

With respect to the α, β basis, the single-particle nuclear spin operators have the matrix form

$$I_x = \tfrac{1}{2}\begin{bmatrix} 0 & 1 \\ 1 & 0 \end{bmatrix} \quad I_y = \tfrac{1}{2}\begin{bmatrix} 0 & -i \\ i & 0 \end{bmatrix} \quad I_z = \tfrac{1}{2}\begin{bmatrix} 1 & 0 \\ 0 & -1 \end{bmatrix} \tag{1.16}$$

The matrix representation of $\mathcal{H}$, then, is

$$\mathcal{H} = \begin{bmatrix} \nu_0\left(1 - \dfrac{\sigma_1}{2} - \dfrac{\sigma_2}{2}\right) + \tfrac{1}{4}J & 0 & 0 & 0 \\[2mm] 0 & \tfrac{1}{2}\nu_0(\sigma_2 - \sigma_1) - \tfrac{1}{4}J & \tfrac{1}{2}J & 0 \\[2mm] 0 & \tfrac{1}{2}J & \tfrac{1}{2}\nu_0(\sigma_1 - \sigma_2) - \tfrac{1}{4}J & 0 \\[2mm] 0 & 0 & 0 & -\nu_0\left(1 - \dfrac{\sigma_1}{2} - \dfrac{\sigma_2}{2}\right) + \tfrac{1}{4}J \end{bmatrix}$$

$$\tag{1.17}$$

where

$$\nu_0 = \frac{\gamma H_0}{2\pi}$$

When this matrix is diagonalized by the use of standard mathematical techniques, one obtains the following eigenvalues and eigenvectors:

$$\nu_0(1 - \tfrac{1}{2}\sigma_1 - \tfrac{1}{2}\sigma_2) + \tfrac{1}{4}J \qquad \alpha\alpha$$

$$-\tfrac{1}{4}J + C \qquad \alpha\beta \cos\theta + \beta\alpha \sin\theta$$

$$-\tfrac{1}{4}J - C \qquad -\alpha\beta \sin\theta + \beta\alpha \cos\theta$$

$$\nu_0(-1 + \tfrac{1}{2}\sigma_1 + \tfrac{1}{2}\sigma_2) + \tfrac{1}{4}J \qquad \beta\beta$$

where

$$C = \tfrac{1}{2}((\nu_0\delta)^2 + J^2)^{1/2} \qquad (1.18)$$

and

$$\tan 2\theta = \frac{J}{\nu_0\delta}$$

Use of Eq. 1.11 leads to the following as allowed transitions:

Transition	Frequency	Relative Intensity
3–1	$\tfrac{1}{2}J + C$	$1 - \sin 2\theta$
4–2	$-\tfrac{1}{2}J + C$	$1 + \sin 2\theta$
2–1	$\tfrac{1}{2}J - C$	$1 + \sin 2\theta$
4–3	$-\tfrac{1}{2}J - C$	$1 - \sin 2\theta$

The analysis of an experimental spectrum consists in inferring from its structure the chemical shifts and spin–spin coupling constants of the nuclei giving rise to the spectrum. From the foregoing, it can be seen that the analysis of an AB spectrum is straightforward; the analysis of more complicated spectra often requires the use of an iterative computer program. One assumes a set of parameters, σ_i and $J_{ii'}$, generates a theoretical spectrum, matches lines between the theoretical and experimental spectra, and iterates to find the set of parameters that will minimize the differences in those lines. Probably the most widely used such program is LAOCN 3, which is available through the Quantum Chemistry Program Exchange, Chemistry Department, Indiana University, Bloomington, Indiana, as QCPE 111.

TIME-DEPENDENT PHENOMENA

There are a number of time-dependent phenomena in NMR spectra. To begin with, we will discuss nuclear magnetic relaxation.

Immediately after a nuclear spin system is put into a magnetic field, the Zeeman energy states of the magnetic moments are equally occupied, which reflects the random orientation of the spins in the absence of a magnetic field. Since alignment of the spins with the field is energetically preferable to alignment opposed to the field (see Eq. 1.2), a redistribution of the spins among the states takes place until the Maxwell–Boltzmann distribution is established. The approach of the system to that state is usually exponential in time, and the time constant of the process is called T_1, the *spin–lattice*, or *longitudinal, relaxation time*. "Spin–lattice" refers to the fact that, for the realignment of spins to take place, there must be an exchange of energy between the spin system and the lattice (containing the rest of the degrees of freedom of the experimental sample). The better the thermal contact between the spin system and the lattice, the shorter T_1 is.

Mathematically, if M_z is the z component of the nuclear magnetization at any time t, and M_0 is the equilibrium magnetization, then typically

$$M_z = M_0(1 - \exp[-t/T_1]) \qquad (1.19)$$

The equilibrium magnetization can be found by using the Maxwell–Boltzmann distribution function. For a system of spin 1/2 nuclei, the energy difference between the two states allowed to the nuclear spin—+ for spin up, – for spin down—is, from Eq. 1.2, just $(2\pi)^{-1}h\gamma H$ and the ratio of nuclei in the higher energy state to those in the lower is

$$\frac{M-}{M+} = \exp(-\Delta E/kT) \qquad (1.20)$$

where T is the temperature and k is Boltzmann's constant.

Under some circumstances, the component of the nuclear magnetization transverse to the magnetic field may be different from zero, its equilibrium value, which corresponds to random orientation. An example of such a circumstance would be immediately after a pulse of rf radiation at the resonance frequency of the nuclei. The approach to the equilibrium value of the transverse component of **M** is called the *spin–spin*, or *transverse, relaxation time*, T_2. Since this relaxation does not involve an interchange of energy with the lattice, T_2 is not necessarily equal to T_1. T_2 is a measure of the length of time spins can maintain phase coherence, if they are initially precessing in phase in the external field, $\mathbf{H}_0$.

Exchange among nuclei in sites with different chemical environments also leads to time-dependent effects in NMR studies. The exchange frequency, or the inverse of the period of stay of a nucleus in a specific site, introduces another time element, which can influence spectra, discussed in detail in Chapter 6.

QUANTUM THEORY OF MAGNETIC RESONANCE PARAMETERS

The parameters determined from the analysis of high-resolution NMR spectra—chemical shifts and spin–spin coupling constants—can be calculated by molecular orbital (MO) or valence bond (VB) theories, because these parameters depend on the molecular electronic configuration. Most recent theories are based on a self-consistent field (SCF) version of MO theory; we will outline the fundamentals of the method in this section.

Molecular Orbital Theory

Schrödinger's equation for a system of N electrons in a molecule can be written

$$\mathcal{H}\Psi = E\Psi \qquad (1.21)$$

where Ψ is the N-electron wave function corresponding to a total energy E, and $\mathcal{H}$ is the Hamiltonian operator for the system. $\mathcal{H}$ is the sum of several parts:

$$\mathcal{H} = \sum_{i=1}^{N} \left(-\frac{\hbar^2}{2m} \nabla_i^2 + V_{ni} \right) + \sum_{i<j} V_{ij} \qquad (1.22)$$

The first term represents the kinetic energy of the electrons, the second term the electric potential energy of the electrons in the field of the nuclei, and the third term the potential energy of Coulomb repulsion of the electrons among themselves.

Even when one makes the approximation that the nuclei are stationary compared to the electrons (the Born–Oppenheimer approximation), one still has an N-body quantum mechanical problem. This calculation is exceedingly complex for any aromatic molecule, so one must look for an approximate scheme. Most developments in the theory of molecular electronic structure have taken the variational principle as the starting point. It can be proved that

$$E_0 \leqq \int \Psi \mathcal{H} \Psi \, d\tau \bigg/ \int \Psi^2 \, d\tau \qquad (1.23)$$

where E_0 is the ground state energy of the system described by the Hamiltonian $\mathcal{H}$, and Ψ is any wave function depending on the coordinates of the particles in the system. One typically uses this principle by assuming a particular form for Ψ that involves one or more parameters and then minimizing the right-hand side of Eq. 1.23 with respect to those parameters.

In the linear-combination-of-atomic-orbitals (LCAO) MO theory, one assumes that the electrons occupy molecular orbitals that are themselves a linear combination of single-particle *atomic* wave functions characteristic of the atoms in the molecule. The physical justification for this assumption is the plausibility of the argument that in the neighborhood of nucleus i, the wave function of an electron should look something like an atomic orbital centered on that nucleus. Mathematically, one writes

$$\psi = \sum_{i=1}^{N} C_i \phi_i \qquad (1.24)$$

where ψ is a molecular orbital, ϕ_i is an atomic orbital centered on nucleus i, and C_i is a coefficient to be treated as a parameter in the variational process.

Using this expansion, one evaluates the right-hand side of Eq. 1.23, and then uses the variational method by minimizing that quantity with respect to each coefficient C_i. When one takes the derivative of the right-hand side with respect to each C_i separately and sets the result equal to zero, one obtains a set of N homogeneous, linear equations for the C_i. A necessary and sufficient condition that a nontrivial solution for the C_i exists is that the determinant of the coefficients vanishes.

Explicitly, if one writes

$$E = \int \psi \mathcal{H} \psi \, d\tau \bigg/ \int \psi^2 \, d\tau \tag{1.25}$$

$$\mathcal{H}_{ij} = \int \phi_i \mathcal{H} \phi_j \, d\tau \tag{1.26}$$

and

$$S_{ij} = \int \phi_i \phi_j \, d\tau \tag{1.27}$$

it can be shown that

$$\sum_i C_i (\mathcal{H}_{ij} - ES_{ij}) = 0 \tag{1.28}$$

leading to

$$\det |\mathcal{H}_{ij} - ES_{ij}| = 0 \tag{1.29}$$

In the lowest order version of the theory, the Hückel MO theory of π electrons, one assumes that $\mathcal{H}_{ij} = 0$ for all pairs of atoms except those that are directly bonded, and that $\mathcal{H}_{ij}$ has the same value (usually called the "resonance integral") for directly bonded neighbors, that $\mathcal{H}_{ii}$, the "Coulomb integral," is the same for all atoms, and that $S_{ij} = \delta_{ij}$, which is equivalent to the assumption that the AO's form an orthonormal set of functions.

The expansion of the determinant in Eq. 1.29 leads to the secular equation, an Nth degree equation with N roots for E. If these roots are labeled $E_J, J = 1, 2, \ldots, N$, then each E_J leads to a set of coefficients, C_{Ji}, and, hence, to an MO

$$\psi_J = \sum_i C_{Ji} \phi_i \tag{1.30}$$

We have, then, a set of N MO's ψ_J corresponding to energies E_J; each MO can accommodate two electrons with opposite spin. In a typical molecule with even N, the $N/2$ lowest energy MO's are doubly occupied.

More complicated MO schemes involve iterative solutions for the ψ_J and E_J, but frequently Hückel MO (HMO) are used as a zeroth order set to begin the iteration process.

The HMO theory incorporates a number of severe approximations, including only nearest-neighbor resonance integrals, assuming that they are all equal, assuming equal Coulomb integrals; and assuming that the atomic orbital basis set is orthogonal (which amounts to assuming that the overlap integral between atomic orbitals on neighboring nuclei is zero). Moreover, it is implicit in the method that the effect on a single electron of all the other electrons in the molecule may be built into the constant resonance integral. An improvement here may be made by going to a self-consistent field (SCF) theory. Certainly the Coulomb interaction between a particular electron and the other electrons will depend on the spatial configuration of the other electrons—that is, the wave functions that the other electrons are occupying. In SCF theory, one assumes an initial configuration of electrons in orbitals (usually, but not necessarily, HMO orbitals), and then calculates a corrected single electron wave function. This new wave function will change the Coulomb interaction energy among the electrons, necessitating a new solution. One continues to iterate until self-consistency is attained—that is, until the wave functions obtained at one stage of the process are sufficiently close to those obtained in the preceding step. We will try to make these ideas more precise mathematically in the following discussion.

The SCF generalization of the HMO basic equation appearing in Eq. 1.29 is as follows:

$$\det |F_{ij} - ES_{ij}| = 0 \tag{1.31}$$

where

$$F_{ij} = \mathcal{H}_{ij}^c + \sum_k \sum_l P_{kl} [(ij|kl) - \tfrac{1}{2}(ik|jl)] \tag{1.32}$$

and

$$S_{ij} = \int \phi_i \phi_j \, d\tau \tag{1.33}$$

In Eq. 1.31, F_{ij} represents the matrix element of the Hamiltonian

$$\mathcal{H} = \sum_\mu \mathcal{H}_\mu + \sum_{\mu < \nu} \frac{e^2}{r_{\mu\nu}} \tag{1.34}$$

between orbitals ϕ_i and ϕ_j:

$$\Gamma_{ij} = \int \phi_i \mathcal{H} \phi_j \, d\tau \tag{1.35}$$

In Eq. 1.34, $\mathcal{H}_\mu$ is the sum of the kinetic energy of the μth electron and its potential energy moving in the electric field of the nuclei, which are assumed to be stationary. The last term in Eq. 1.34 is the Coulomb interaction energy among the electrons themselves. The factor P_{kl} in Eq. 1.32 is the usual bond order

$$P_{ij} = 2 \sum_J C_{Ji} C_{Jj} \tag{1.36}$$

where the summation is over occupied molecular orbitals. The other factors entering in Eq. 1.32 are defined as follows:

$$(ij|kl) = \int \int \phi_i(1)\,\phi_j(1)\,\frac{e^2}{r_{12}}\,\phi_k(2)\,\phi_l(2)\,d\tau_1 d\tau_2 \tag{1.37}$$

and are double integrals involving the coordinates of electrons one and two and the AO ϕ_i, ϕ_j, ϕ_k, and ϕ_l. This integral arises in the SCF MO calculation when the Hamiltonian in Eq. 1.34 is evaluated with respect to the total N electron wavefunction Ψ, which is a function of single electron MO's, ψ_J, which are themselves functions of atomic orbitals ϕ_i as defined in Eq. 1.30. The derivation of Eq. 1.32 may be found in numerous references. (See the list at the end of this chapter.)

It is clear that a solution of Eq. 1.31 must involve iteration, because of the presence of P_{ij} in the definition of F_{ij}. One cannot evaluate P_{ij} to use in Eq. 1.31 without having expressions for C_{Ji}, as follows from Eq. 1.36. The SCF iteration scheme, then, involves assuming initial values for the C_{Ji} (for example, the solutions coming from an HMO calculation), and then solving for an improved set of C_{Ji} by expanding the determinant in Eq. 1.31.

Now consider the integrals $\mathcal{H}_{ij}^c$, and the integrals defined in Eq. 1.37. If one starts with an explicit formula for the AO, such as Slater orbitals, the mathematical evaluation of the integrals involved can be quite difficult. Some widely used versions of SCF theory employ different integral approximations. Several of the most widely used of these theories incorporate some empirically determined parameters in assigning values to these integrals. The different approximations vary primarily in their treatment of differential overlap. The overlap integral is defined by

$$S_{ij} = \int \phi_i \phi_j \, d\tau \tag{1.38}$$

In the HMO theory, as was pointed out earlier, S_{ij} is taken to be one if i equals j, and is zero otherwise. The zero differential overlap (ZDO) approximation is obtained by assuming that the product $\phi_i \phi_j$ is always zero for all i and all j if i is different from j. This is, of course, an even stronger approximation than the analogous one in HMO theory: the ZDO approximation not only results in

S_{ij} being equal to zero if i is different from j, but also in many of the integrals of Eq. 1.32 being equal to zero. Indeed, there is only one nonzero electron repulsion integral in the ZDO approximation, and that is

$$(ii|jj) = \gamma_{ij} \tag{1.39}$$

One can also define a resonance integral by the expression

$$\int \phi_i \mathcal{H}^c \phi_j \, d\tau = \beta_{ij} \tag{1.40}$$

In terms of these, one can write for the diagonal elements of the F matrix in the ZDO approximation

$$F_{ii} = \mathcal{H}_{ii}^c + \sum_j \gamma_{ij} P_{ij} - \tfrac{1}{2} \gamma_{ii} P_{ii} \tag{1.41}$$

and for the off diagonal elements

$$F_{ij} = \beta_{ij} - \tfrac{1}{2} P_{ij} \gamma_{ij} \tag{1.42}$$

How are the values for γ_{ij} and β_{ij} chosen? The ZDO–SCF theory was applied to π-electron systems by Pariser, Parr, and Pople by choosing γ_{ii} to be the experimentally measured difference between the ionization potential and the electron affinity of a carbon atom in the $\pi(sp^2)_3$ valence state. Since an empirical quantity appears in an otherwise theoretically derived expression, the method is said to be "semi-empirical." A number of different mathematical relationships between γ_{ii} and γ_{ij} on the one hand and β_{ij} and S_{ij} on the other have been suggested in the literature.

The ZDO–SCF semi-empirical method has been applied to π-electron systems in various approximate forms with considerable success.

More recently the semi-empirical SCF–MO method has been extended from a treatment of π-electron systems alone to include all valence electrons. One widely used approximation is the CNDO (complete neglect of differential overlap) method. In this method all molecular integrals $(ij|kl)$ are neglected unless $i = j$ and $k = l$. Moreover, integrals of the form $(ii|jj)$ are taken to be the same for all valence orbitals i on one atom and j on another atom. The most widely used form of this approximation is the CNDO/2 form, which uses a particular set of empirical parameters in the assignment of values for the integrals involved in the SCF calculation.

A significant improvement over the CNDO/2 method is called the INDO (intermediate neglect of differential overlap) approximation. In the INDO method, all one-center exchange integrals of the form $(ij|ij)$, where both orbitals are on the same atom, are included. The INDO extension of CNDO/2 makes possible the theoretical interpretation of phenomena that depend upon the exchange interaction. As we will see in the chapter on NMR spin–spin coupling, the π-electron contribution to H–H coupling constants provides an example of such a phenomenon. In general, INDO calculations are in better agreement with experiment than CNDO/2.

A somewhat different theoretical approach is used in the MINDO (modified intermediate neglect of differential overlap) SCF method. In the MINDO approximation, the parameterization was designed to get the best agreement with experimental data, rather than to reproduce *ab initio* SCF calculations with the same AO, as in the case of INDO.

Calculations of NMR Parameters

MO theory can be used to calculate NMR chemical shifts and spin–spin coupling constants. In this section, we will outline briefly the general *theoretical procedures* involved, and in later chapters we will provide the details of the calculations.

Magnetic shielding arises from the secondary local magnetic field set up at the site of the nucleus by electronic currents induced in the molecule by turning on the magnetic field necessary to perform the NMR experiment. With a reasonably good theory of molecular electronic structure, quantum mechanics enables calculation of the induced electronic currents, and hence the local magnetic field. Two different theoretical approaches have been used in attacking this problem— the "test-dipole" and "current-density" approaches.

In the test-dipole approach, one writes down a Hamiltonian for the molecular electronic system and includes a term representing the interaction of the electrons with a magnetic vector potential arising from the external magnetic field and a magnetic dipole at the site of the nucleus concerned. One then solves Schrödinger's equation for the energy of the molecular electronic system. There will be a term in this energy that is proportional to the strength of the nuclear magnetic dipole, μ. Since the coupling of a magnetic dipole, μ, with a magnetic field, $\mathbf{H}'$, is

$$E = -\mu \cdot \mathbf{H}'$$

one identifies the coefficient of μ in the energy of the electron system with $-\mathbf{H}'$, the local field at the site of μ. Generally, $\mathbf{H}'$ will turn out to be proportional to $\mathbf{H}_0$, the strength of the external magnetic field, so one can identify the coefficient of $-\mathbf{H}_0$ in the expression for $\mathbf{H}'$ with the shielding parameter σ, as implied in Eq. 1.6.

In the current density approach, one evaluates the quantum mechanical current density for the electrons by solving Schrödinger's equation in the presence of the external magnetic field, $\mathbf{H}_0$, and then forming the current density operator as usually defined in quantum mechanics:

$$\mathbf{j} = \frac{e\hbar}{2im} [\psi^* \nabla \psi - (\nabla \psi^*) \psi] \tag{1.43}$$

Knowing the current density, one can calculate the local magnetic field produced by that current density with the classical equation

$$\mathbf{H}' = \int \frac{\mathbf{j} \times \mathbf{r}}{r^3} \, d\tau \tag{1.44}$$

Again, the shielding parameter σ is identified with the coefficient of $-\mathbf{H}_0$ in the expression for $\mathbf{H}'$.

Both test-dipole and current-density calculations of chemical shielding parameters have been used extensively in the past, and results are comparable; indeed, in many cases, they can be shown to be equivalent.

In the calculation of spin–spin coupling constants, one first recalls that the physical interpretation of spin–spin coupling is that it arises from an indirect coupling between two nuclear spins by way of an electron spin. To calculate the spin–spin coupling constant, J, one begins by writing down a Hamiltonian for the electron system in the molecule, including specifically the Fermi contact hyperfine interaction between the electrons and the nuclei. This interaction, which was first postulated to explain hyperfine splitting in atomic spectra, represents an interaction between two magnetic dipoles at the same point in space. Again, as in the case of the calculation of shielding parameters, there are two ways to proceed. In the first case, Schrödinger's equation is solved in some approximation to obtain wave functions for the electron system in the absence of the Fermi interaction. Time-independent perturbation theory is then used to obtain the change in energy levels caused by the presence of the Fermi interaction. The form of the perturbed energy will involve terms that include the interaction of an electron spin with two nuclei. These terms, bilinear in the two nuclear spins, are proportional to the scalar product of the two spins. Since this is the same form as the spin–spin coupling interaction, the coefficient of the scalar product (which will depend on the details of the molecular electron system) is identified as the spin–spin coupling constant. In the other approach, Schrödinger's equation is solved in the presence of the perturbing nuclear spins, and the same identification is made in the final expression for the energy of the electron system. More recently, the latter method, in the finite perturbation form, has been the more widely used. The two methods give similar results when similar approximate forms for calculating the solution to Schrödinger's equation are used.

Relaxation times T_1 and T_2 may also be calculated theoretically. In the standard approach, the Hamiltonian for the system containing the nuclear spins is partitioned into three parts—first, the Zeeman energy of the nuclear magnetic moments in the external magnetic field; second, a part incorporating the energy of the "bath" or "lattice" in which the nuclear spin system finds itself (for example, in a system of liquid water this term would include all the translational, rotational, and vibrational degrees of freedom of the water molecules);

and third, a part representing the coupling between the spin system and the lattice. This third term will contain coordinates of both the spin system and the bath, and is a measure of the degree of thermal contact between the two systems. A typical example of the third term would be the magnetic dipole–dipole interaction between nuclear spins in the same molecule. This interaction has the form

$$\mathcal{H}_{dd} = \frac{\mu_1 \cdot \mu_2}{r^3} - \frac{3(\mu_1 \cdot \mathbf{r})(\mu_2 \cdot \mathbf{r})}{r^5} \tag{1.45}$$

The nuclear magnetic moments are clearly coordinates of the spin system; the radius vector $\mathbf{r}$ is a coordinate of the bath and will be changing in direction as the molecule rotates in the liquid.

The third, coupling, term in the Hamiltonian of the system is treated as a time-dependent perturbation that induces transitions between the Zeeman levels of the spin system. The stronger the coupling, the greater the transition probability caused by the interaction, and the shorter the relaxation time. In an NMR experiment the relaxation mechanisms are acting in a fashion to establish the Boltzmann equilibrium distribution of spins among the various energy states. If resonance is occurring, transitions that disturb that equilibrium distribution of spins, in competition with the relaxation mechanism, will be taking place. Relaxation times may vary over many orders of magnitude. Relaxation mechanisms other than the dipole–dipole interaction include the interaction of a nuclear electric quadrupole moment (which is related to the spin coordinates) with the electric field gradient at the site of the nucleus, the spin–rotation interaction, and others.

SELECTED REFERENCE BOOKS

Nuclear Magnetic Resonance

A. Abragam, *The Principles of Nuclear Magnetism*, Oxford, Fairlawn, N.J., 1961.

E. D. Becker, *High Resolution NMR: Theory and Chemical Applications*, 2nd ed., Academic, New York, 1980.

A. Carrington and A. D. McLachlan, *Introduction to Magnetic Resonance with Applications to Chemistry and Chemical Physics*, Harper and Row, New York, 1967.

W. T. Dixon, *Theory and Interpretation of Magnetic Resonance Spectra*, Plenum, New York, 1972.

J. W. Emsley, J. Feeney, and L. H. Sutcliffe, *High Resolution Nuclear Magnetic Resonance Spectroscopy*, Pergamon, New York, 1965–1966.

T. C. Farrar and E. D. Becker, *Pulse and Fourier Transform NMR: Introduction to Theory and Methods*, Academic, New York, 1971.

J. D. Memory, *Quantum Theory of Magnetic Resonance Parameters*, McGraw-Hill, New York, 1968.

W. W. Paudler, *Nuclear Magnetic Resonance*, Allyn and Bacon, Boston, 1971.

J. A. Pople, W. G. Schneider, and H. J. Bernstein, *High Resolution Nuclear Magnetic Resonance*, McGraw-Hill, New York, 1959.

C. P. Slichter, *Principles of Magnetic Resonance*, Springer-Verlag, New York, 1978.

Molecular Orbital Theory

W. T. Borden, *Modern Molecular Orbital Theory for Organic Chemists*, Prentice-Hall, Englewood Cliffs, N.J., 1975.

M. J. S. Dewar, *The PMO Theory of Organic Chemistry*, Plenum, New York, 1975.

H. F. Hameka, *Advanced Quantum Chemistry*, Addison-Wesley, Reading, Mass., 1965.

J. N. Murrell and A. J. Harget, *Semi-Empirical Self-Consistent-Field Molecular Orbital Theory of Molecules*, Wiley, New York, 1972.

R. G. Parr, *The Quantum Theory of Molecular Electronic Structure*, Benjamin, New York, 1963.

J. A. Pople and D. L. Beveridge, *Approximate Molecular Orbital Theory*, McGraw-Hill, New York, 1970.

L. Salem, *The Molecular Orbital Theory of Conjugated Systems*, Benjamin, New York, 1966.

G. A. Segal, *Semiempirical Methods of Electronic Structure Calculation*, Plenum, New York, 1977.

A. Streitwieser, Jr., *Molecular Orbital Theory for Organic Chemists*, Wiley, New York, 1961.

Two

Theory of Chemical Shifts in Aromatic Compounds

In this chapter, we discuss the theory of chemical shifts in aromatic molecules. Experimental results are tabulated in the following chapter, and comparison with different theories relating chemical shifts to electronic structure is made.

PROTON CHEMICAL SHIFTS

Introduction

Protons in the vicinity of an aromatic ring exhibit chemical shifts to higher frequencies relative to protons in an otherwise similar chemical environment.[1] Specifically, the proton chemical shift in benzene is about 1.5 ppm from the proton chemical shift in ethylene.[2] Moreover, the greater the number of rings near a proton, the greater the shift; for example, the proton resonances of benzene are at δ_H 7.27, whereas those of anthracene are shifted at δ_H 7.39 to 8.36.[3]

It has long been assumed that the chemical shift of a particular nucleus can be reasonably partitioned into diamagnetic and paramagnetic contributions from electrons associated with that nucleus, a sum of local contributions from other atoms, and finally a "nonlocal" contribution caused by mobile π electrons.[4] Since the bond hybridizations in benzene and ethylene are similar, the first two contributions should be much the same and the observed differences should be caused by the last two terms. Early efforts to account for the chemical shifts in aromatic hydrocarbons singled out the last contribution, in the form of a "ring-current" effect,[5] as the dominant factor. Some recent evidence[6] indicates that the third contribution, from electrons on neighboring atoms, also plays a significant role. We discuss these two effects separately.

The Ring-Current Effect

The considerations described in the preceding section lead one to think that the chemical shift of a proton near an aromatic ring depends at least in part on the secondary magnetic field produced by currents in the π-electron system of the ring—currents induced by turning on the magnetic field necessary for the NMR experiment. These nonlocal currents about the ring periphery are traditionally called "ring currents." Haigh and Mallion have recently published an excellent review on the subject.[7] Indeed, it has been suggested that the aromaticity of a ring be defined in terms of its capacity to sustain ring currents (references 8–12). Alternatively, Musher has suggested that ring currents may be a fiction arising from approximations in the molecular orbital (MO) theory of chemical shifts and diamagnetic anisotropies in aromatic compounds, and has sought to explain the effects on the basis of a local current alone (references 13 and 14). While Musher's work along this line serves as a useful warning against too uncritical an acceptance of a model, agreement between his theory and experimentation has not been so successful as that of SCF–MO ring-current theory.[15] In addition, the signs of the outer and inner proton chemical shifts in the annulenes (to be discussed in the following chapter) support a model in which ring currents play a role, rather than an exclusively local model.[16]

Ring currents were first hypothesized by Pauling[17] to account for the large diamagnetic anisotropy in the magnetic susceptibility of single crystals of polycyclic aromatic hydrocarbons. The susceptibility measured normal to the plane of the molecules is considerably greater than that in the plane. Pauling proposed that this anisotropy is caused by π-electron ring currents induced by turning on the magnetic field necessary to perform the susceptibility experiment. Pauling's calculations, based on an electric network model, were in sufficiently good agreement with experimental results to produce some confidence in the model.

In the 1950s, NMR spectra of moderate resolution were obtained by Bernstein, Schneider, and Pople (BSP) for several aromatic hydrocarbons.[5] They proposed a simple semiclassical model of the effect of ring currents on chemical shifts that accounted reasonably well for the observed facts.

This calculation (in Gaussian units) goes as follows: it is assumed that the six π-electrons in benzene are free to circulate about the ring under the influence of the induced emf produced when the external magnetic field H_0 is turned on. According to the Larmor precession theorem,[18] they will circulate about the ring with angular frequency

$$\omega = \frac{eH_0}{2mc} \tag{2.1}$$

where e and m are the electron's charge and mass, respectively, and c is the velocity of light.

A charge, e, moving in a circle with an angular frequency, ω, represents a current

$$I = \frac{e\omega}{2\pi} \tag{2.2}$$

so the six electrons produce a net current of

$$I = \frac{3e^2 H_0}{2\pi mc} \tag{2.3}$$

It is clear from Figure 2.1 that this current will give rise to a deshielding field at the peripheral site of the proton; that is, the induced field will tend to reinforce the external magnetic field. The magnitude of that secondary field, and hence the contribution to the shielding, can be estimated by replacing the current loop with a magnetic dipole at the center of the ring having the same dipole moment as the current loop itself. That dipole moment will be

$$\mu = \frac{I(\pi a^2)}{c} \tag{2.4}$$

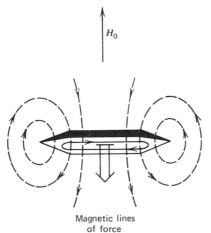

Magnetic lines
of force

Figure 2.1. Current and magnetic lines of force induced in benzene by a primary field, H_0. From J. A. Pople, W. G. Schneider, and H. J. Bernstein, *High-Resolution Nuclear Magnetic Resonance*, McGraw-Hill, New York (1959).

where a is the radius of the ring. The field at a proton that is at a distance r from the center of the ring is simply

$$H' = \frac{\mu}{r^3} = \frac{3e^2 a^2 H_0}{2mc^2 r^3} \tag{2.5}$$

The chemical shift is the negative of the coefficient of H_0 in this expression.

Applied magnetic fields along any two mutually perpendicular axes *in* the plane of the ring will give rise to no such currents, so the π-electron contribution to those two components of the shielding tensor will be zero. The molecule rotates rapidly and randomly through the liquid in a typical high-resolution NMR experiment: to compare the calculated with the observed chemical shift we must average the three diagonal values of the shielding tensor. The result for the π-electron chemical shift in benzene according to this simple model is

$$\sigma_\pi = -\frac{e^2 a^2}{2mc^2 r^3} \tag{2.6}$$

Inserting appropriate values for a and r, we obtain 1.75×10^{-6} for the ring current contribution to the proton chemical shift to benzene. This result agrees both in direction (deshielding) and roughly in magnitude with the observed effects of aromaticity on proton chemical shifts.

Bernstein, Schneider, and Pople extended this theory by calculating chemical shifts for several polycyclic aromatic hydrocarbons, assuming that the current in each ring of such a polycyclic compound was equal to the benzene current, and that the magnetic effects of the current loops could be approximated adequately by the magnetic field of point dipole. The agreement with experimental results was encouraging.

The BSP theory was subsequently refined by Waugh and Fessenden[2] and by Johnson and Bovey,[19] who replaced the point dipole with a distributed π-electron ring current. A classical calculation was performed to determine the local magnetic field caused by currents distributed around the rings. More recently, Farnum and Wilcox[20] have extended the method by treating the current as distributed over toroidal shells. Theories using the Biot–Savart theorem from classical electromagnetic theory have been developed by Longuet-Higgins,[21] Salem,[22] and Haddon.[23]

Quantum mechanical theories of the ring-current effect are based on various levels of approximation of molecular orbital (MO) theory. The initial calculations were based on Hückel MO theory,[24,25] while the more recent calculations have been based on some form of a self-consistent field (SCF) method (references 26–31). These theories have been of two types—either a "current-density" or a "test-dipole" approach. In the first current-density calculation, Pople[24] calculated the quantum mechanical current density, as defined through Schrö-dinger's equation, caused by external magnetic field.

At approximately the same time, McWeeny[25] published the first test-dipole calculation of ring-current chemical shifts. In this theoretical approach, the effect of a point nuclear magnetic dipole is included in the Hamiltonian for the π-electron system. The energy of the system is then expanded in a series involving the strength of the nuclear magnetic dipole, μ. Since the energy of coupling of a dipole with a magnetic field H is of the form $-\mu \cdot \mathbf{H}$, the coefficient of $-\mu$ in the total energy of the system is identified with the local magnetic field caused by the circulating π-electron currents at the site of the nucleus. More precisely, since the chemical shift is defined by $H' = -\sigma H$, where H' is the induced field, we write

$$\sigma = -\frac{\partial^2 E}{\partial \mu \partial H} \tag{2.7}$$

The McWeeny theory is a generalization of the Hückel molecular orbital (HMO) method to the situation when the electron system is in the presence of a magnetic field. The magnetic field can be described by a vector potential, $\mathbf{A}$, where the field itself can be found from $\mathbf{A}$ by

$$\mathbf{H} = \text{curl } \mathbf{A} = \nabla \times \mathbf{A}$$

If the Hamiltonian for a system in the absence of a magnetic field is given by

$$\mathcal{H} = \sum_i \left(\frac{p_i^2}{2m}\right) + V \tag{2.8}$$

then it can be shown that the Hamiltonian of the same system in a magnetic field described by vector potential $\mathbf{A}$ is

$$\mathcal{H} = \sum_i \frac{1}{2m}\left(\mathbf{p}_i + \frac{e}{c}\mathbf{A}_i\right)^2 + V \tag{2.9}$$

where $\mathbf{A}_i$ is the vector potential seen by the ith electron.

There is the question of the choice of the "gauge" when a vector potential is introduced; since the curl of the gradient of any scalar equals zero, if one defines a new vector potential, $\mathbf{A}'$, by

$$\mathbf{A}' = \mathbf{A} + \nabla \zeta \tag{2.10}$$

then $\mathbf{A}'$ describes the same magnetic field, since

$$\text{curl } (\nabla \zeta) = 0 \tag{2.11}$$

for any ζ. This property of the vector potential is known as "gauge invariance." Moreover, it can be shown that if Ψ is a solution to Schrödinger's equation with a choice of gauge $\mathbf{A}$, then a new solution to Schrödinger's equation with gauge defined by Eq. 2.10 is

$$\Psi' = \Psi \exp\left(-\frac{ie\zeta}{\hbar c}\right) \tag{2.12}$$

These considerations led London to define the so-called "gauge invariant atomic orbitals," or GIAO, by

$$\phi_i = \phi_i^{(0)} \exp\left(-\frac{ie}{\hbar c}\mathbf{A}_i \cdot \mathbf{r}\right) \tag{2.13}$$

where $\phi_i^{(0)}$ is the zero field atomic orbital, and $\mathbf{A}_i$ is the vector potential at the site of nucleus i. It can be shown then[18] that

$$\left(\frac{\hbar}{i}\nabla + \frac{e}{c}\mathbf{A}\right)^2 \phi_i^{(0)} \exp\left(-\frac{ie}{\hbar c}\mathbf{A}_i \cdot \mathbf{r}\right)$$

$$= \exp\left(-\frac{ie}{\hbar c}\mathbf{A}_i \cdot \mathbf{r}\right)\left[\frac{\hbar}{i}\nabla + \frac{e}{c}(\mathbf{A} - \mathbf{A}_i)\right]^2 \phi_i^{(0)} \tag{2.14}$$

Note that the GIAO depends only on the "local" vector potential $(\mathbf{A} - \mathbf{A}_i)$, which is zero at the center of atom i.

One now modifies the HMO procedure (see Chapter 1):

$$|\mathcal{H}_{ij} - E\delta_{ij}| = 0 \tag{2.15}$$

where

$$\mathcal{H}_{ij} = \int \phi_i^* \mathcal{H} \phi_j \, d\tau \tag{2.16}$$

but where the ϕ_i are now GIAO, as defined in Eq. 2.13.

Substituting from Eq. 2.13, we obtain

$$\mathcal{H}_{ij} = \int \exp\left[\frac{ie}{\hbar c}(\mathbf{A}_i - \mathbf{A}_j) \cdot \mathbf{r}\right] \phi_i^{(0)*}$$

$$\times \left\{\frac{1}{2m}\left[\mathbf{p} + \frac{e}{c}(\mathbf{A} - \mathbf{A}_j)\right]^2 + V\right\}\phi_j^{(0)} \, d\tau \tag{2.17}$$

We will neglect the local vector potential $(\mathbf{A} - \mathbf{A}_i)$ appearing in Eq. 2.17 for two reasons: first, since GIAO have been chosen, the factor $(\mathbf{A} - \mathbf{A}_i)$ is zero at the center of the orbital, and should be small in comparison with the momentum term in that region; and second, any current effects caused by the local vector potential will correspond to local currents, not the interatomic ring currents we are looking for.

At this point, McWeeny, following London in his theory of diamagnetic susceptibility in aromatic compounds, made the approximation of replacing $\mathbf{r}$ in Eq. 2.17 by its value at the midpoint of the i–j bond:

$$\mathbf{r} \longrightarrow \tfrac{1}{2}(\mathbf{R}_i + \mathbf{R}_j) \tag{2.18}$$

where $\mathbf{R}_i$ is the distance from the origin (the site of the nuclear dipole μ) to the center of atom i. This approximation leads to considerable mathematical simplification and has been made plausible by saying that the maximum orbital overlap occurs at that point.

With this approximation, one obtains

$$\beta_{ij} \equiv \mathcal{H}_{ij}$$
$$= \beta_{ij}^{(0)} \exp\left[\frac{ie}{2\hbar c}(\mathbf{A}_i - \mathbf{A}_j) \cdot (\mathbf{R}_i + \mathbf{R}_j)\right] \tag{2.19}$$

where $\beta_{ij}^{(0)}$ is the usual zero-field Hückel resonance integral.

We must now turn to the specific form for the vector potential $\mathbf{A}$. McWeeny considered the external magnetic field $\mathbf{H}_0$ in addition to a magnetic field caused by a nuclear magnetic moment at the site of the nucleus whose chemical shift was to be determined. It can be shown directly, by taking the curl of $\mathbf{A}$, that a choice for $\mathbf{A}$ giving such a magnetic field is

$$\mathbf{A} = -\tfrac{1}{2}H_0\mathbf{r} \times \mathbf{k} + \mu\mathbf{r} \times \mathbf{k}/r^3 \tag{2.20}$$

where $\mathbf{k}$ is a unit vector normal to the molecular plane (taken to be along the z axis).

One then carries out the HMO procedure, treating the magnetic effects as a perturbation. Having obtained the perturbed energy of the electron system, one identifies the coefficient of $-\mu$ in the energy of the electron system with the local magnetic field, H', at the site of that dipole moment, in accordance with Eq. 2.7, which leads to the following expression for H' (references 18 and 25):

$$H' = 2\beta\left(\frac{e}{\hbar c}\right)^2 \frac{S^2 H}{a^3}(\sigma_1 + \sigma_2) \tag{2.21}$$

where

$$\sigma_1 = \sum_{(ij)} P_{ij} s_{ij}^2 k_{ij} \tag{2.22}$$

and

$$\sigma_2 = \sum_{(ij)} \sum_{(kl)} \beta\pi_{(ij)(kl)} s_{ij} s_{kl} \frac{k_{ij} + k_{kl}}{2} \tag{2.23}$$

In Eqs. 2.22 and 2.23, P_{ij} is the HMO order of the ij bond, and $\bar{\pi}_{(ij)(kl)}$ is the imaginary part of the complex polarizability of the ij and kl bonds. Specifically,

$$P_{ij} = \sum_J \lambda_J C_{Ji} C_{Jj}^* \tag{2.24}$$

and

$$\bar{\pi}_{ij,(kl)} = \pi_{ij,kl} - \pi_{ij,lk} + \pi_{kl,ij} - \pi_{lk,ij} \tag{2.25}$$

where

$$\pi_{ij,kl} = \sum_J \lambda_J \sum_{K(\neq J)} \frac{C_{Ji}^* C_{Kj} C_{Kk}^* C_{Jl}}{E_J - E_K} \tag{2.26}$$

C_{Ji} is the coefficient of the ith atomic orbital in the Jth molecular orbital, E_J is the energy of that MO, and λ_J is 0, 1, or 2, the occupation number of the Jth MO; s_{ij} is the signed area, in units of S, the area of a benzene ring, of the triangle with vertices at the nuclear magnetic moment and the centers of atoms i and j; k_{ij} is defined by

$$k_{ij} = a^3 (R_i^{-3} + R_j^{-3})$$

where a is the C–C bond length.

The McWeeny theory has been the most widely used quantum mechanical method of ring-current chemical shifts in aromatic compounds. Detailed comparisons with experimental results appear in the following chapter.

Both the Pople[24] and McWeeny[25] theories are based on HMO and make use of the London approximation, in which one replaces $\mathbf{r}$ in a molecular integral involving two 2-p atomic orbitals (AO) by its value at the center of the bond between the two carbons on which the AO are centered, as was mentioned earlier. Even though the agreement between these theories and experiments for planar protons in polycyclic aromatics is reasonably good, approximations made in the theories were such that the method was open to criticism.

The natural extension of the McWeeny theory to incorporate self-consistent field molecular orbitals in place of HMO has been carried out by different authors, using two different degrees of approximation. The principal distinction between these two methods involves whether the SCF procedure was "coupled" or "uncoupled." The potential energy of the electron system in the SCF method is a function of the MO (see Chapter 1). When the magnetic field is introduced theoretically by the replacing of $\mathbf{p}$ by $\mathbf{p} + e\mathbf{A}/c$, a change in the MO, which is reflected in a change in V, results. The theoretical method that takes this dependence of V on $\mathbf{A}$ into account is called the "coupled" Hartree–Fock method. This form of the theory was used by Hall and Hardisson[26,27] in their calculations of ring-current shielding. In the "uncoupled" approach, however, for which justification has been given by Amos and Roberts,[32] the potential is taken to be a function only of the field-unperturbed orbitals. An analysis of the error introduced by this approximation (after using the Feenberg–Goldhammer[32-35] "geometric" correction) has been discussed by a number of authors (see also the footnote on p. 328 of reference 7). In the SCF extensions the ring-current concept was still shown to have validity, and the current-density and point-dipole methods were shown to be equivalent.

Amos and Roberts[28] began by writing down the quantum mechanical current-density for the jth electron:

$$\mathbf{J}_j(r_j) = \int \left[\frac{\hbar e i}{2m} (\Psi \nabla_j \Psi^* - \Psi^* \nabla_j \Psi) - \frac{e^2}{mc} \mathbf{A}_j \Psi^* \Psi \right] d\tau_1 \cdots d\tau_{2N} \quad (2.27)$$

In this expression Ψ is the wave function for the $2N$ electron system, and the integral excludes the space coordinates of electron j. They then assumed that Ψ could be represented as a single Slater determinant of N molecular orbitals, each occupied by two electrons with opposite spin. The usual vector potential for a constant magnetic field is:

$$\mathbf{A}_j = \tfrac{1}{2} H \mathbf{k} \times \mathbf{r}_j \quad (2.28)$$

where $\mathbf{r}_j$ is the vector position of the jth electron relative to the origin of the vector potential, and $\mathbf{k}$ is the unit vector in the z direction, assumed to be the direction of the magnetic field, $\mathbf{H}$, necessary for the NMR experiment. They further assumed that the ϕ_i, the molecular orbitals in the presence of the magnetic field, could be expanded in a power series as follows:

$$\phi_i = \phi_i^0 + iH\phi_i' + H^2 \phi_i'' + \cdots \quad (2.29)$$

where ϕ_i^0 is the molecular orbital in the absence of a magnetic field, and ϕ_i' is real. They also expanded $\mathbf{J}(\mathbf{r})$ as a power series in H. The zeroth order term for $\mathbf{J}$, corresponding to the current in the absence of a magnetic field, vanishes identically, as it should on physical grounds. The first-order term has the form

$$\mathbf{J}'(\mathbf{r}) = 2 \sum_{i=1}^{N} \left\{ -\frac{\hbar e}{m} [\phi_i' \nabla \phi_i^0 - \phi_i^0 \nabla \phi_i'] - \frac{e^2}{2mc} \mathbf{k} \times \mathbf{r} \phi_i^0 \phi_i^0 \right\} \quad (2.30)$$

From this expression one can separate the local current density from the non-local current density of π orbitals. The zero field π orbitals are expressed in the usual LCAO method as a linear combination of atomic $2p_z$ orbitals ω_r:

$$\phi_i^0 = \sum_{r=1}^{m} a_{ir}^0 \omega_r \quad (2.31)$$

One can then write the new molecular orbitals in the presence of the magnetic field as follows:

$$\phi_i = \sum_r a_{ir} \chi_r \tag{2.32}$$

where the coefficients a_{ir} are given by

$$a_{ir} = a_{ir}^0 + \frac{ie}{\hbar c} H a'_{ir} + \cdots \tag{2.33}$$

and

$$\chi_s = \omega_s \exp\left(-\frac{ie}{\hbar c} \mathbf{A}_s \cdot \mathbf{r}\right) \tag{2.34}$$

The χ_s are the so-called "gauge invariant atomic orbitals" originally introduced by London.

From the last three equations, it follows that

$$\phi'_i = \frac{e}{\hbar c} \sum_s a'_{is} \omega_s - \frac{e}{2\hbar c} \sum_s a_{is}^0 \mathbf{k} \times \mathbf{R}_s \cdot \mathbf{r} \omega_s \tag{2.35}$$

The expression for the nonlocal current density then becomes

$$\begin{aligned} \mathbf{J}'_\pi(\mathbf{r}) = \frac{e^2}{2mc} \sum_{s,t=1}^{M} & [P'_{st} - \tfrac{1}{2} P_{st}^0 \,\mathbf{k} \times (\mathbf{R}_s - \mathbf{R}_t) \cdot \mathbf{r}]\, (\omega_s \nabla \omega_t - \omega_t \nabla \omega_s) \\ & - \tfrac{1}{2} P_{st}^0 \,\mathbf{k} \times (2\mathbf{r} - \mathbf{R}_s - \mathbf{R}_t)\, \omega_s \omega_t \end{aligned} \tag{2.36}$$

where

$$\begin{aligned} P_{st} &= 2 \sum_{i=1}^{P} a_{is}^0 a_{it}^0 \\ P'_{st} &= 2 \sum_{i=1}^{p} (a'_{is} a_{it}^0 - a_{is}^0 a'_{it}) \end{aligned} \tag{2.37}$$

and the sum is over occupied π-molecular orbitals. The calculation of the coefficients a_{is} is done using SCF theory.

Amos and Roberts then argued, again following London, that the largest contribution to terms of the form $\mathbf{r}\omega_s\omega_t$ occurs at the midpoint of the s–t bond where $\mathbf{r} = 1/2(\mathbf{R}_s + \mathbf{R}_t)$, and wrote

$$\mathbf{J}_\pi(\mathbf{r}) \approx \frac{e^2}{2mc} \sum_{s,t=1}^{M} C_{st}(\omega_s \nabla \omega_t - \omega_t \nabla \omega_s) \tag{2.38}$$

where

$$C_{st} = P'_{st} - \tfrac{1}{2} \mathbf{k} \times \mathbf{R}_s \cdot \mathbf{R}_t P^0_{st} \tag{2.39}$$

To evaluate the chemical shift caused by this current density, we need only use the standard formula for the magnetic field, $\mathbf{F}'$, caused by a current density, $\mathbf{J}'_\pi$:

$$\mathbf{F}' = \frac{1}{c} \int \frac{\mathbf{r} \times \mathbf{J}'_\pi}{r^3} \, d\mathbf{r} \tag{2.40}$$

where the origin of $\mathbf{r}$ is taken to be the site of the proton whose chemical shift is to be determined.

This can be written

$$\mathbf{F}' = -\frac{e^2}{2mc^2} \sum_{s,t=1}^{M} C_{st} \mathbf{F}'_{st} \tag{2.41}$$

where

$$\mathbf{F}'_{st} = \int \frac{\mathbf{r} \times (\omega_s \nabla \omega_t - \omega_t \nabla \omega_s)}{r^3} \, d\mathbf{r} \tag{2.42}$$

The coefficients C_{st} (Eq. 2.20) are calculated by SCF-MO theory. The least approximate method for calculating these SCF coefficients is to use the "coupled" procedure employed by Hall and Hardisson.[26,27] Roberts points out, however, that judicious use of the Hückel method is reasonably satisfactory for benzenoid hydrocarbons, and that a method of intermediate approximation is to use the uncoupled procedure described by Amos and Musher.[33-35]

Once molecular orbital coefficients C_{st} have been determined, the integral F'_{st} defined in Eq. 2.42 must be evaluated. Even using the London approximation

$$\mathbf{r} \longrightarrow \tfrac{1}{2} (\mathbf{R}_s + \mathbf{R}_t) \tag{2.43}$$

there are several alternatives as to how the $1/r^3$ factor should be approximated. Two suggestions were made by Amos and Roberts and calculations based on each of the two versions of the London approximation were compared with experiments. The first, called LAI, is

$$\frac{1}{r^3} \longrightarrow \left| \frac{1}{2} (\mathbf{R}_s + \mathbf{R}_t) \right|^{-3} \tag{2.44}$$

The second, called LAII, is

$$\frac{1}{r^3} \longrightarrow \frac{1}{2} (R_s^{-3} + R_t^{-3}) \tag{2.45}$$

Calculations based on LAI were in better agreement with experimental results than those based on LAII.

However, Roberts[29] evaluated the F'_{st} as they stand, using numerical methods, and has concluded that little error is introduced if one uses the London approximation for bonds more than four bond lengths from the nucleus whose chemical shift is to be determined. A theoretical investigation of the effects of the London approximation appears in work by Coulson, Gomes, and Mallion.[30]

Memory[36] has developed a formula that uses LAI to determine the chemical shift as a function of position with respect to an aromatic ring. In the same paper, ring-current chemical shift calculations were made in which the F'_{st} were evaluated exactly; that is, without making the LAI approximation. These two calculations based on LAI and the exact evaluation of F'_{st} are compared with ring-current calculations by Haigh and Mallion[37] using the McWeeny[25] test-dipole method, and by Johnson and Bovey[19] using the semiclassical method described earlier. These results appear as Figures 2.2 and 2.3.

The corrections from inclusion of overlap integrals between neighboring atoms have been calculated by a number of authors.[38-41]

There have been several calculations of ring currents in aromatic compounds using the free electron theory rather than molecular orbital theory.[42-47] In free electron theory, the delocalized π electrons are assumed to be confined by a constant potential, provided by the other electrons and the nuclei, to ring-shaped regions about the periphery of aromatic rings. Turning on the magnetic field necessary for a susceptibility experiment or an NMR experiment induces

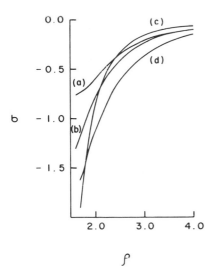

Figure 2.2. σ as a function of ρ for points in the plane of the ring. σ is in parts per million, and ρ in units of the C–C bond length. (*a*) The "exact" calculation; (*b*) the LAI calculation; (*c*) based on the McWeeny theory; (*d*) based on the Johnson–Bovey result (see reference 36).

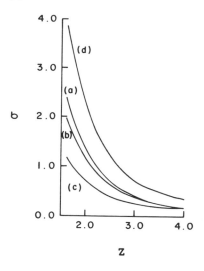

Figure 2.3. σ as a function of z for points above the ring along its axis of symmetry. σ is in parts per million, and z in units of the C–C bond length. (*a*) The "exact" calculation; (*b*) the LAI calculation; (*c*) based on the McWeeny theory; (*d*) based on the Johnson–Bovey result (see reference 36).

currents in these rings in accordance with Faraday's law on induction. The work done in this field has been reviewed by Walnut.[47]

The Neighbor Anisotropy Effect

If we consider the chemical shielding of a proton to be the result of the four contributions described earlier,[4] we need a way to estimate the magnitude of the third term, local contributions from other atoms, for protons in aromatic compounds. This effect is usually estimated by assuming that point magnetic dipoles of magnitudes related to the magnetic susceptibility of the atom exist at the other atoms in the molecule, and calculating contributions to the local field at the site of a proton for such a distribution of magnetic dipoles.[48,49] The contribution from a particular atom that is a neighbor of a proton has an interesting property: it is zero unless the susceptibility is anisotropic. Figure 2.4 shows an atom, X, in the neighborhood of a proton, H, for two orientations of the HX axis with respect to the external magnetic field, $\mathbf{H_0}$. It is clear that for one orientation the effect of the electron currents at atom X produces a *shielding* effect at H, and for another orientation a *deshielding* effect. The molecule in a high-resolution NMR experiment will be rotating rapidly and randomly, so an average of the shielding tensor is measured. The details of the analysis, as given below, indicate that if the susceptibility of atom X is isotropic in direction, the shielding and deshielding effects will cancel, so the net contribution to the chemical shift will be proportional to any anisotropy in the susceptibility— the difference between the susceptibility perpendicular to the XH direction and parallel to it.

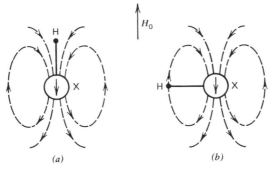

Figure 2.4. Secondary magnetic field caused by the diamagnetism of neighboring atom. (*a*) Primary field parallel to XH; (*b*) primary field perpendicular to XH. Broken lines represent magnetic lines of force. From J. A. Pople, W. G. Schneider, and H. J. Bernstein, *High-Resolution Nuclear Magnetic Resonance*, McGraw-Hill, New York (1959).

The magnetic dipole moment μ induced by the field $\mathbf{H}_0$ when $\mathbf{H}_0$ is parallel to the XH bond is, from the definition of the susceptibility,

$$\mu^{\|} = \chi^{\|} H_0 \tag{2.46}$$

Correspondingly, for $\mathbf{H}_0$ perpendicular to the XH bond,

$$\mu^{\perp} = \chi^{\perp} H_0 \tag{2.47}$$

The magnetic field set up by a magnetic moment, μ, at a point at a distance, R, along its axis is

$$H'_{\|} = 2R^{-3}\mu^{\|} \tag{2.48}$$

At right angles to the axis, the field is

$$H'_{\perp} = -R^{-3}\mu^{\perp} \tag{2.49}$$

Using $H' = -\sigma H_0$, we have

$$\sigma^{\|} = -2R^{-3}\chi^{\|} \tag{2.50}$$

and

$$\sigma^{\perp} = R^{-3}\chi^{\perp} \tag{2.51}$$

If we assume axial symmetry about the bond so that both transverse components of the susceptibility are equal, then the observed chemical shift, in which an average is taken over all molecular orientations, is

$$\sigma = -\tfrac{1}{3}R^{-3}(2\chi^{\|} - \chi^{\perp} - \chi') \tag{2.52}$$

so there is a nonzero value for such a shift because of a neighbor only if $\chi^\| \neq \chi^\perp$ — that is, only if the susceptibility is anisotropic.

Barfield, Grant, and Ikenberry[6] have made use of recent determinations of the components of the shielding tensor for ^{13}C in solid benzene[50] to estimate the magnitude of the neighbor anisotropy shielding of protons in benzenoid hydrocarbons. Their approach was as follows: they took the experimental values for the components of the ^{13}C shielding tensor (two of which, σ_{22}^C and σ_{33}^C, are equal) and used as a model to account for this shielding a local electron current, I, induced by the magnetic field, flowing in a circular orbit of radius a. They then used the expression for the magnetic field at a certain distance from a current loop and arrived at the following formula for the contribution to the shielding of a proton averaged over all directions as the molecule rotates through the liquid:

$$\langle \sigma^H \rangle = \left\{ \frac{1}{3} a^3 \sigma_{22}^C (a^2 + r_{CH}^2)^{-3/2} + \frac{a(\sigma_{11}^C + \sigma_{22}^C)}{\pi(a + r_{CH})} \left[K + \frac{a + r_{CH}}{a - r_{CH}} E \right] \right\} \quad (2.53)$$

In this expression σ_{11}^C and σ_{22}^C are the components of the ^{13}C shielding tensor, r_{CH} is the distance from the carbon to the hydrogen in question, and K and E are complete elliptic integrals with argument

$$k^2 = 4a\rho \left[(a + \rho)^2 + z^2 \right]^{-1} \quad (2.54)$$

where ρ and z are the elliptical coordinates of the proton. There will be one such term for each carbon in the molecule for the neighbor anisotropy contribution to σ for a proton in an aromatic compound.

Barfield, Grant, and Ikenberry[6] conclude from their results that the neighbor anisotropy effect contributes significantly to the deshielding of protons in aromatic compounds, and is comparable in importance to the ring-current effect. They further consider in some detail the shielding of (4n) and (4n + 2) annulenes, and find that the experimental results are consistent with a model in which both local anisotropy and ring-current effects are important.

Vogler[51-53] has included both ring-current and local anisotropy contributions in a theory he has used to calculate proton chemical shifts in a variety of aromatic compounds. The method is applicable to atoms with either sp^2 or sp hybridization. The sigma core is approximated by localized two-center bonds, and the pi system is treated using extended coupled Hartree–Fock perturbation theory. The method is successful as applied to both benzenoid aromatic hydrocarbons and annulenes.

Explicitly, the expression for the chemical shift is

$$\sigma = \sigma^{RC} + \sigma^{LA} + \sigma_\mu^0 + \sigma_\nu^q \quad (2.55)$$

where the ring-current contribution is

$$\sigma^{RC} = -f^{\sigma\pi}(E_{11}^{z,n} + E_{011}^{z,n})/3 \qquad (2.56)$$

and the local anisotropy term is

$$\sigma^{LA} = \sigma_{011}^{LA} + \sigma_{011}^{LA} \qquad (2.57)$$

where

$$\sigma_{11}^{LA} = -E_{11}'/3 \qquad (2.58)$$

and

$$\sigma_{011}^{LA} = -(f^{\delta\sigma}E_{011}^{l,\sigma} + E_{011}^{x,y,1})/3 \qquad (2.59)$$

In Eq. 2.55, σ_μ^0 simply defines the zero of the chemical shift scale, and

$$\sigma_\nu^q = a_\nu \Delta q_\nu \qquad (2.60)$$

gives the dependence of the shielding parameter on the excess π-electron charge density, Δq_ν, which is important in a consideration of aromatic heterocyclics.[54]

In Eq. 2.56, the quantities designated E are contributions to the second-order perturbed energy of the electron system, and involve molecular integrals with respect to the atomic orbitals appearing in a coupled Hartree–Fock self-consistent field procedure. The parameters labeled f are chosen to provide an optimum fit to experimental data on chemical shifts of benzenoid hydrocarbons. Details of the calculation appear in reference 51.

CHEMICAL SHIFTS OF NUCLEI OTHER THAN HYDROGEN

Introduction

Earlier in the chapter, we discussed how the chemical shift of a particular nucleus is the sum of diamagnetic and paramagnetic contributions from the electrons associated with that particular atom, contributions from local currents on other atoms, and contributions from delocalized currents.[4] For the case of protons, the differences observed in chemical shifts are caused primarily by the last two contributions, owing to the similarity of the immediate electron environment of protons in different sites. To calculate chemical shifts for other nuclei, we must take into account the large variation in the contribution from electrons associated with that particular atom, since these may vary significantly from one molecular site to another.

A good starting point for describing the local electron contribution to the chemical shift is the theory of chemical shielding developed by Ramsey.[55] Ramsey's theory is a "test-dipole" method, in which he considers the effect on the energy of the electron system of the external magnetic field and the magnetic field of the test dipole at the site of the nucleus. Using second-order perturbation theory, Ramsey obtained an expression for the change in energy level of the electron system caused by the presence of the nuclear magnetic dipole. Using an argument analogous to that given earlier in this chapter, he identified the shielding parameter with the coefficient of $-\mu H$ in the perturbed energy. His final expression for the zz component of the shielding parameter tensor (taking μ and $\mathbf{H}$ in the z direction) is

$$\sigma_{zz} = \frac{e^2}{2mc^2} \int \frac{x^2 + y^2}{r^3}\, \rho\, d\tau + \frac{e^2 \hbar^2}{m^2 c^2 \Delta E} \left(0 \left| \sum_{\nu\nu'} r_\nu^{-3} \frac{\partial^2}{\partial\phi_\nu \partial\phi_{\nu'}} \right| 0\right) \quad (2.61)$$

The first term is called the diamagnetic contribution and the second, the paramagnetic contribution, for reasons that will be specified later. The derivation of Eq. 2.61 may be found in reference 18 or 55.

In Eq. 2.61, $\rho = |\Psi|^2$, where Ψ is the electron wave function; the matrix element in the second term is with respect to the ground state of the electron system, the sum is over electron pairs, and ΔE is an average of the excitation energies of the electron system.

The first term in Ramsey's equation had been derived earlier by Lamb,[56] who had assumed a spherically symmetrical electron wave function in an atom and that the effect of the external magnetic field would cause Larmor precession of the electron cloud. He then used classical electromagnetic theory to calculate the secondary induced field at the atomic center. The shielding for the case of a spherically symmetrical electron distribution at its center is always diamagnetic, so the first term in Ramsey's formula is called the diamagnetic shielding term. The second term may be regarded as a correction of Lamb's formula that takes into account the deviation from spherical symmetry of the electron system. Typically, this contribution will be opposite in sign from the first term, and hence is called the paramagnetic contribution to the chemical shift.

Ramsey's formula does not provide an accurate method for calculating chemical shifts in large molecules, because in this case the calculated chemical shift is the small difference of two large terms, so that small errors in both terms are magnified in the final result. However, when Ramsey's formula is applied only to the electrons associated with a particular nucleus in a molecule, it can be useful.

Chemical Shifts of Fluorine

The chemical shifts of fluorine in different molecular sites illustrate the importance of the local contributions to the chemical shift. The fluoride ion is spherically symmetrical, so the second term in Ramsey's equation is negligibly small in comparison with the first. In covalently bonded fluorine, on the other hand, there is considerable deviation from spherical symmetry because of the presence of p electrons; in F_2, for example, the paramagnetic term is large and negative, with the result that there is 630 ppm difference in the chemical shielding of the fluorine nucleus in F_2 relative to that observed in HF. The fluorine resonance in the molecule UF_6 is over 900 ppm below that in HF. In view of this deviation, it is not surprising that the chemical shift of the fluorine nucleus in a wide variety of sites is nearly linearly related to the electronegativity of the atom to which it is bonded.[57,58]

Saika and Slichter[4] made use of this line of thought by calculating the chemical shift difference of fluorine in F_2 and ionic fluorine by calculating the observed shift as resulting from the paramagnetic contribution from the p electrons in covalent fluorine. Their result was

$$\Delta \sigma = -\frac{2}{3} \left(\frac{e^2 \hbar^2}{m^2 c^2} \right) \left\langle \frac{1}{r^3} \right\rangle_{2p} \frac{1}{\Delta E} \tag{2.62}$$

In this expression, r^{-3} is averaged over the $2p$ electrons.

This approach was generalized by Karplus and Das[59], who used MO's in the Ramsey formula to calculate fluorine chemical shifts. The result of their calculation for the paramagnetic contribution to the chemical shift of a fluorine atom directly bonded to a single other atom is

$$\sigma = \sigma_0 |1 - s - I + \tfrac{1}{2}(4 - p_{xx} - p_{yy})| \tag{2.63}$$

In this equation σ_0 is the Saika–Slichter difference between ionic and covalent fluorine shielding, I is the ionic character of the bond in question, s is the degree of sp hybridization and p_{xx} and p_{yy} are the populations of the p_{xx} and p_{yy} orbitals. Note that the larger the ionic character of the bond, the smaller σ, in accord with the earlier discussion in this section.

An extension of the work of Karplus and Das[59] by Prosser and Goodman[60] relaxes the restriction of considering only electrons localized on the fluorine to include the effect of the π-electron distribution in the molecule as a whole.

Dewar and Kelemen[61] extended the Karplus–Das–Prosser–Goodman work to include the long-range effects of electrons on atoms distant from the fluorine atom in question. They developed the equation

$$\sigma = aq_i + b \sum_{(m \neq i)} \frac{q_m}{r_m^3} \tag{2.64}$$

where a and b are constants, q_i is the charge density on the fluorine, and q_m is the charge density on atom m distant r_m from the fluorine nucleus.

Chemical Shifts of Carbon

It has been observed experimentally that the chemical shifts of ^{13}C nuclei in aromatic compounds, relative to the ^{13}C benzene chemical shift, are proportional to the excess π-electron charge density on the carbon.[62]

$$\Delta\sigma_A = \alpha(\rho_A - 1) \tag{2.65}$$

In this expression ρ_A is the π-electron density on the carbon in question.

Karplus and Pople[62] developed theoretical considerations that account for this dependence. They first concluded, with order of magnitude calculations, that the local diamagnetic contribution, and contributions from local currents on other atoms and interatomic ring currents, were not large enough to account for the observed differences in chemical shifts. This finding turned their attention to the paramagnetic local contribution, which Saika and Slichter[4] and Karplus and Das[59] had earlier shown was the key to understanding differences in chemical shifts of ^{19}F nuclei in a variety of related molecules. Karplus and Pople then used an LCAO-MO theory in the Ramsey formula for the paramagnetic local term (see Eq. 2.61). For a carbon atom bonded to three other carbon atoms, their expression for this term is

$$\sigma_p^{AA} = -[e^2\hbar^2/(2m^2c^2\Delta E)]\langle r^{-3}\rangle_{2p} \times [2 + \tfrac{4}{9}(P_{z_A z_B} + P_{z_A z_C} + P_{z_A z_D})]. \tag{2.66}$$

and for a carbon atom bonded to two other carbon atoms and a hydrogen they obtained

$$\sigma_p^{AA} = -[e^2\hbar^2/(2m^2c^2\Delta E)]\langle r^{-3}\rangle_{2p}$$
$$\times [2 + \tfrac{4}{9}\lambda_H(1 - P_{z_A z_A}) + \tfrac{4}{9}(P_{z_A z_B} + P_{z_A z_C})]. \tag{2.67}$$

In these expressions, r^{-3} is evaluated with respect to the $2p$ electrons on the carbon, ΔE is an average MO excitation energy, and $P_{\mu\nu}$ is the bond order as defined in Chapter 1 for the μ and ν orbitals in the system. λ_H in Eq. 2.67 is a polarity parameter, which is small but not necessarily equal to zero.

The authors then proceeded to show that the factor $\langle r^{-3}\rangle_{2p}$ leads to a linear proportionality of the term to the π-electron charge density. The physical reason for this result is that as electron density on a carbon atom increases, the $2p$ orbitals will expand because of electron repulsion, and $\langle r^{-3}\rangle_{2p}$ will become correspondingly smaller. Using Slater orbitals one can show that

$$\langle r^{-3}\rangle_{2p} = \tfrac{1}{24}(Z_A/a_0)^3 \tag{2.68}$$

where Z_A is the effective nuclear charge. For Slater orbitals, Z_A is 3.25 for a neutral carbon atom, plus .35 additional screening per $2p$ electron, so that

$$Z_A = 3.25 - 0.35(\rho_A - 1) \tag{2.69}$$

ρ_A will be small in comparison with 1, so we may make the expansion

$$(1 - \rho_A)^3 \approx 1 - 3\rho_A \tag{2.70}$$

When Eqs. 2.68, 2.69, and 2.70 are used with Eqs. 2.66 and 2.67, one finds a linear relationship between the chemical shielding parameter and ρ_A, the excess charge density.

The equation predicting the linear relationship between the ^{13}C chemical shift and the excess in the π-electron charge density has been widely tested,[63] as will be reported in Chapter 5. The charge density itself has been calculated in numerous different levels of approximation, using the different versions of semi-empirical molecular orbital theory described in Chapter 1.

In addition to the Karplus–Pople theory, the Karplus–Das expression originally derived for fluorine has also been tested experimentally as applied to ^{13}C chemical shifts.[64,65] In some instances, a better correlation with experimental chemical shifts has been obtained with total $\sigma + \pi$ electron density rather than π-electron density alone.

REFERENCES

1. C. W. Haigh, R. B. Mallion, and E. A. G. Armour, *Mol. Phys.* **18**, 751 (1970).

2. J. S. Waugh and R. W. Fessenden, *J. Am. Chem. Soc.* **79**, (1957).

3. C. W. Haigh and R. B. Mallion, *Mol. Phys.* **18**, 737 (1970).

4. A. Saika and C. P. Slichter, *J. Chem. Phys.* **22**, 26 (1954).

5. H. J. Bernstein, W. G. Schneider, and J. A. Pople, *Proc. R. Soc.* **A236**, 515 (1956).

6. M. Barfield, D. M. Grant, and D. Ikenberry, *J. Am. Chem. Soc.* **97**, 6956 (1975).

7. C. W. Haigh and R. B. Mallion, *Prog. NMR Spectrosc.* **13**, 303 (1980).

8. H. P. Figeys, *Tetrahedron Lett.* **38**, 4625 (1966).

9. R. C. Fahey and G. C. Graham, *J. Chem. Phys.* **69**, 4417 (1968).

10. R. H. Martin, N. Defay, F. Geerts-Evrard, and S. Delavarenne, *Tetrahedron* **20**, 1073 (1964).

11. R. J. Batterham, L. Tsai, and H. Ziffer, *Aust. J. Chem.* **17**, 163 (1964).

12. T. J. Batterham, L. Tsai, and H. Ziffer, *Aust. J. Chem.* **18**, 1959 (1965).

13. J. I. Musher, *J. Chem. Phys.* **43**, 4081 (1965).

14. J. I. Musher, *Adv. Magn. Resonance* **2**, 177 (1966).

15. J. Nowakowski, *Theor. Chim. Acta* **10**, 79 (1968).

16. J. M. Gaidis and R. West, *J. Chem. Phys.* **46**, 1218 (1967).

17. L. Pauling, *J. Chem. Phys.* **4**, 673 (1936).

18. J. D. Memory, *Quantum Theory of Magnetic Resonance Parameters*, McGraw-Hill, New York, 1968.

19. C. E. Johnson, Jr., and F. A. Bovey, *J. Chem. Phys.* **29**, 1012 (1958).

20. D. G. Farnum and C. F. Wilcox, *J. Am. Chem. Soc.* **89**, 5379 (1967).

21. H. C. Longuet-Higgins and L. Salem, *Proc. R. Soc.* **A257**, 445 (1960).

22. L. Salem, *Molecular-Orbital Theory of Conjugated Systems*, Chap. 4, Benjamin, New York, 1966.

23. R. C. Haddon, *Tetrahedron* (a) **28**, 3613 (1972); (b) **28**, 3635 (1972).

24. J. A. Pople, *Mol. Phys.* **1**, 175 (1958).

25. R. McWeeny, *Mol. Phys.* **1**, 311 (1958).

26. G. G. Hall and A. Hardisson, *Proc. R. Soc.* **A268**, 328 (1962).

27. G. G. Hall, A. Hardisson, and L. M. Jackman, *Tetrahedron* **19** Supp. **2**, 101 (1963).

28. A. T. Amos and H. G. Ff. Roberts, *Mol. Phys.* **20**, 1073, 1081, 1089 (1971).

29. H. G. Ff. Roberts, *Mol. Phys.* **27**, 843 (1974).

30. C. A. Coulson, J. A. N. F. Gomes, and R. B. Mallion, *Mol. Phys.* **30**, 713 (1975).

31. E. R. Long and J. D. Memory, *J. Chem. Phys.* **61**, 3865 (1974).

32. A. T. Amos and H. G. Ff. Roberts, *J. Chem. Phys.* **50**, 2375 (1969).

33. (a) E. Feenberg, *Phys. Rev.* **103**, 1116 (1956); (b) P. Goldhammer and E. Feenberg, *ibid.*, **101**, 1233 (1956).

34. (a) A. T. Amos and J. I. Musher, *Mol. Phys.* **13**, 509 (1967); (b) J. M. Schulman and J. I. Musher, *J. Chem. Phys.* **49**, 4845 (1968).

35. A. T. Amos, *J. Chem. Phys.* **52**, 603 (1970).

36. J. D. Memory, *J. Magn. Resonance* **27**, 241 (1977).

37. C. W. Haigh and R. B. Mallion, *Org. Magn. Resonance* **4**, 203 (1972).

38. H. Brooks, *J. Chem. Phys.* **9**, 463 (1961).

39. G. W. Wheland and S. L. Matlow, *Proc. Natl. Acad. Sci. USA* **38**, 364 (1952).

40. L. Caralp and J. Hoarau, *J. Chem. Phys.* **65**, 1565 (1968).

41. M. Mayot, G. Berthier, and B. Pullman, *J. Phys. Radium* **12**, 717 (1951).

42. J. R. Platt, *J. Chem. Phys.* **17**, 484 (1949).

43. K. Ruedenberg and C. W. Scherr, *J. Chem. Phys.* **21**, 1965 (1953).

44. A. D. McLachlan and M. R. Baker, *Mol. Phys.* **4**, 255 (1961).

45. T. K. Rebane, *Vestn. Leningradskova Univ.*, (a) *1957* **10**, p. 11; (b) *1957* **16**, p. 19; (c) *1957* **22**, p. 70.

46. T. K. Rebane, *Zh. Eksp. Teor. Fiz.* **38**, 963 (1960) (Soviet Physics—JETP, **11**, 694 (1960).

47. T. H. Walnut, (a) *Chem. Phys. Lett.* **33**, 956 (1975); (b) *J. Chem. Phys.* **64**, 4531 (1976).

48. J. A. Pople, *Proc. R. Soc.* **A239**, 550 (1957).

49. H. M. McConnell, *J. Chem. Phys.* **27**, 226 (1957).

50. A. Pines, M. G. Gibby, and J. S. Waugh, *Chem. Phys. Lett.* **15**, 373 (1972).

51. H. Vogler, *J. Am. Chem. Soc.* **100**, 7464 (1978).

52. H. Vogler, *J. Mol. Struct.* **51**, 289 (1979).

53. H. Vogler, *Tetrahedron* **35**, 657 (1979).

54. T. B. Cobb and J. D. Memory, *J. Chem. Phys.* **50**, 4262 (1969).

55. N. F. Ramsey, *Phys. Rev.* **78**, 699 (1950).

56. W. E. Lamb, *Phys. Rev.* **60**, 817 (1941).

57. J. W. Emsley, J. Feeney, and L. H. Sutcliffe, *High Resolution Nuclear Magnetic Resonance Spectroscopy*, Pergamon, Oxford, 1965, p. 151.

58. J. N. Shoolery and H. E. Weaver, *Annu. Rev. Phys. Chem.* **6**, 443 (1955).

59. M. Karplus and T. P. Das, *J. Chem. Phys.* **34**, 1683 (1961).
60. F. Prosser and L. Goodman, *J. Chem. Phys.* **38**, 374 (1963).
61. M. J. S. Dewar and J. Kelemen, *J. Chem. Phys.* **49**, 499 (1968).
62. M. Karplus and J. A. Pople, *J. Chem. Phys.* **38**, 2803 (1963).
63. K. A. K. Ebraheem and G. A. Webb, *Prog. NMR Spectrosc.* **11**, 149 (1977).
64. T. D. Alger, D. M. Grant, and E. G. Paul, *J. Am. Chem. Soc.* **88**, 5397 (1966).
65. A. J. Jones, T. D. Alger, D. M. Grant, and W. M. Litchman, *J. Am. Chem. Soc.* **92**, 2386 (1970).

Three

Experimental Chemical Shifts of ^1H and ^{19}F

In this chapter we describe some experimental determinations of ^1H and ^{19}F chemical shifts in aromatic compounds. Where it is possible, these experimental determinations are examined in light of the theoretical predictions.

PROTON CHEMICAL SHIFTS

Benzenoid Hydrocarbons

A broad and systematic study of the high-resolution proton NMR spectra of polycyclic aromatic hydrocarbons has been carried out by Haigh and Mallion.[1] Their work, an extension of considerable prior experimental effort by a number of others,[2-13] resulted in complete analyses of the spectra of 15 of these compounds. Intermolecular effects were minimized by obtaining spectra at various concentrations and then extrapolating each parameter to "infinite dilution."

For the simple spectra of relatively symmetrical molecules, analyses in terms of chemical shifts and spin–spin coupling constants were possible from 60-MHz spectra. For more complicated spectra, experimental results at 100 MHz and 220 MHz were obtained. Inter-ring spin–spin coupling is quite small in these compounds,[14,15] so the spectra can be resolved into the superposition of contributions no more complicated than the ABCD type. For example, naphthalene, 2, has an AA′BB′ spectrum. Iterative computer methods were quite satisfactory in effecting complete analyses of all the spectra.

As an example of the spectrum analysis technique, we will consider benzan-thracene, **8** (see Figure 3.1). Guided by the fact that inter-ring spin–spin coupling is negligible, we would expect single lines from protons 9 and 10, an AB quartet for protons 3 and 4, and two ABCD multiplets, one each for the proton sets 1, 2, 3, 4 and 5, 6, 7, 8. The peaks resulting from protons 9 and 10 are clearly observable; the peak at a larger chemical shift is associated with the 9 proton, since it is closer, on the average, to more rings than proton 10. All chemical shift theories predict that next to proton 9, the largest chemical shift will be for proton 1, because of its central position close to many rings. The multiplet around 1920, therefore, is associated with proton 1. An AB quartet is observable in the complicated part of the spectrum at a smaller chemical shift and is identified with the spectrum arising from protons 3 and 4. Proceeding in this fashion, one

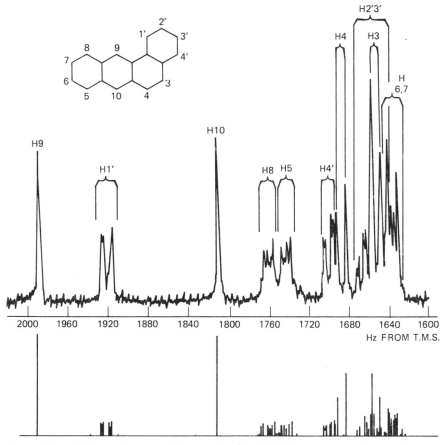

Figure 3.1. 1,2-Benzanthracene (2.4 percent w/v): 220-MHz spectrum with stick diagram. (From reference 1.)

can make tentative zeroth order assignments of the chemical shifts and coupling constants for the system; use of an iterative spectrum analysis program (see Chapter 1) leads to a least-squares fit of the experimental spectrum. The stick spectrum representing the optimum parameter set for fitting the experimental results appears below the experimental spectrum.

The proton chemical shifts inferred are shown in Table 3.1; spin–spin coupling constants are considered in Chapter 5.

Having obtained experimental values for chemical shifts in a variety of benzenoid hydrocarbons, Haigh, Mallion, and Armour[16] compared those values with theoretical predictions based on McWeeny's theory[17] (see Chapter 2). Specifically, they used Eq. 2.21; the question of what value to use for the resonance integral β arose immediately. Some standard values for β give absolute magnitudes of σ not in good agreement with experiments.[18] To eliminate the necessity of assigning a value to β, they decided, as a number of other authors had,[5,6] to test the ratio of the theoretical chemical shift of a particular proton to the theoretical chemical shift of the proton in benzene. For a comparison with experiments, the question of what value to use for the experimental chemical shift in benzene was left open. Authors of earlier works[5,6] had taken the value 1.55 ppm as an estimate for the ring-current contribution to the chemical shift of the proton in benzene, as had been suggested in the early 1960s by Spiesecke and Schneider,[19] or they used the benzene shift with respect to a shift in a compound in which ring currents did not occur—for example, cyclohexa-1, 3-diene, cycloocta-1, 3, 5-triene, and cyclooctatetraene.[20] Haigh, Mallion, and Armour[16] improved on these procedures by simply plotting the theoretical value $H'/H'_{benzene}$ as a function of experimentally observed chemical shifts. For the

Table 3.1.   Observed Proton Chemical Shifts (δ_H) for Polycyclic Aromatic Hydrocarbons in CCl_4

Compound	Proton	δ_H
Benzene (1)		7.27
Naphthalene (2)	1	7.73
	2	7.38
Anthracene (3)	1	7.93
	2	7.39
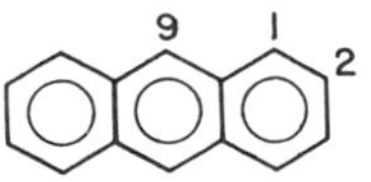	9	8.36

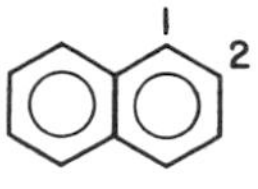

Table 3.1. (Continued)

Compound	Proton	δ_H
Phenanthrene	1	7.80
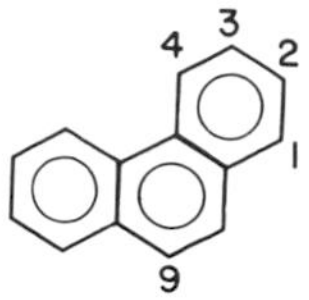	2	7.51
	3	7.57
	4	8.62
	9	7.65

(4)

Compound	Proton	δ_H
Chrysene	1	8.66
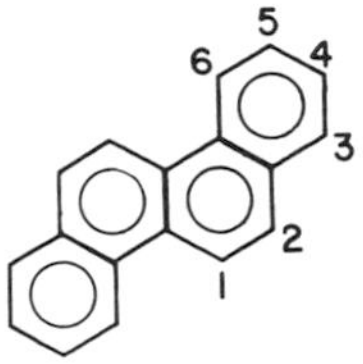	2	7.93
	3	7.90
	4	7.52
	5	7.62
	6	8.72

(5)

Compound	Proton	δ_H
Triphenylene	1	8.61
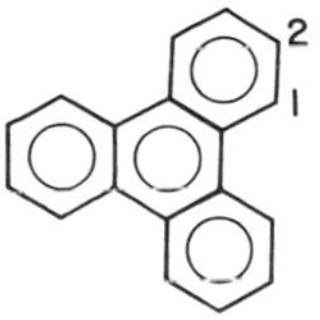	2	9.58

(6)

Compound	Proton	δ_H
Pyrene	1	8.00
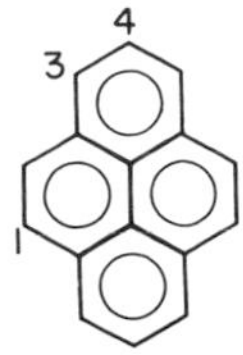	3	8.10
	4	7.93

(7)

Compound	Proton	δ_H
1,2-Benzanthracene	1'	8.77
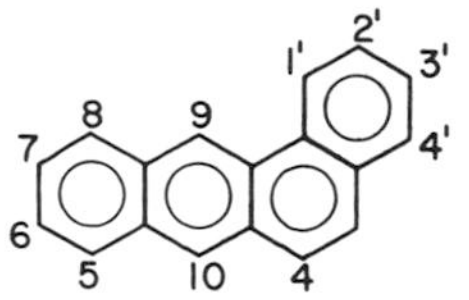	2'	7.59
	3'	7.53
	4'	7.76
	3	7.55
	4	7.72
	5	7.95
	6	7.47
	7	7.47
	8	8.03
	9	9.91
	10	8.28

(8)

Source. Reference 1.

majority of the protons—those that meet the criteria of being "not overcrowded" (see later discussion)—they were able to fit the line by standard regression analysis with the following equation:

$$\delta_{obs} = 1.56(H'/H'_{benzene}) + 5.66 \qquad (3.1)$$

The correlation coefficient is 0.96; these results are plotted (using the $\tau = 10 - \delta$ convention) in Figure 3.2. It is to be noted that Eq. 3.1 gives a ring-current contribution to the chemical shift of the benzene proton of 1.56 ppm, quite close to earlier estimates.

Recently, Mallion[21] has compared the experimental data with the prediction of the simple Pople, Schneider, and Bernstein "point-dipole" model discussed in Chapter 2. He finds agreement with experiments to be as good as that with more theoretically refined methods, provided the ring-current intensities are calculated by the McWeeny HMO method. The agreement is better than for the recent "π-bond" model.[22,23]

Not all of the proton experimental chemical shifts observed appear in the plot shown in Figure 3.2—specifically, protons that are known to lie some-

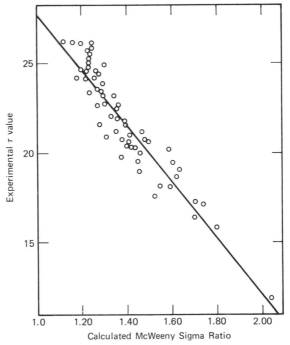

Figure 3.2. The regression line (Eq. 3.1) and plot of observed τ value versus sigma ratio, calculated from the McWeeny theory. (From reference 1.)

what outside the molecular plane—as, for example, the 4 and 5 protons in phenanthrene, **4**, for which steric interference exists. Such protons for which steric interference results in a significant displacement from the molecular plane have been called "crowded" or "hindered" protons. The protons whose shifts appear in Figure 3.2 are those that are planar, such as protons in benzene, **1**, naphthalene, **2**, anthracene, **3**, and so forth. It has been observed by a number of workers[4-6,24,25] that crowded protons have chemical shifts that, perhaps expectedly, deviate from the predictions of a theory of ring-current shift based on planar molecules. Memory and Cobb[6] observed that the effects of crowding could be largely removed by adding 0.54 ppm to the calculated ring-current shift downfield. Haigh, Mallion, and Armour[16] observed that the average discrepancy for singly overcrowded protons (leaving out such doubly crowded protons as the 9 proton in 1,2;7,8-dibenzanthracene, **14**) is 0.62 ppm, with a range of values from 0.44 ppm to 0.76 ppm.

Reid[24] originally suggested that the shift of the crowded protons in phenanthrene, **4**, arose from a van der Waals interaction. Bartle and Smith[25] noted that for phenanthrene the neighbor anisotropy shifts (see Chapter 2) for the carbon–carbon and carbon–hydrogen bonds have opposite signs for the uncrowded protons, but the same sign for crowded protons. This finding could explain the low field shift of the crowded protons. Haigh, Mallion, and Armour[16] tentatively concluded that the effect was not large enough to explain the observed shift, and they favored a bond-polarization mechanism proposed by Cheney.[26] Further discussion of the crowded proton chemical shifts in connection with the Barfield, Grant, and Ikenberry[27] calculations is given below.

Roberts[18] has carried out calculations of proton chemical shifts in these compounds using the ring-current theory based on the uncoupled version of SCF theory (see Chapter 2), and these appear in column II of Table 3.2. The "four-

Table 3.2. Proton Chemical Shifts

Compound	H_f	I^a	II^b	III^c	Experimental
Benzene, **1**	1				0
Naphthalene, **2**	1	0.54	0.48	0.44	0.46
	2	0.16	0.21	0.18	0.11
Anthracene, **3**	1	0.66	0.70	0.52	0.66
	2	0.22	0.28	0.24	0.12
	9	1.09	0.93	0.89	1.09
Phenanthrene, **4**	1	0.63	0.63	0.51	0.53
	2	0.23	0.30	0.24	0.24
	3	0.28	0.23	0.27	0.30

Table 3.2. (Continued)

Compound	H_f	I[a]	II[b]	III[c]	Experimental
	4	1.19	0.67	0.89	1.35
	9	0.71	0.67	0.58	0.38
Chrysene, **5**	1	1.36	0.90		1.39
	2	0.82	0.69		0.66
	3	0.68	0.75		0.63
	4	0.31	0.41		0.25
	5	0.31	0.33		0.35
	6	1.27	0.85		1.45
Triphenylene, **6**	1	1.28	0.73	0.95	1.34
	2	0.33	0.36	0.28	0.31
Pyrene, **7**	1	0.77	0.88	0.71	0.73
	3	0.70	0.79	0.72	0.83
	4	0.33	0.56	0.40	0.66
1,2-Benzanthracene, **8**	1'	1.33	0.68	1.00	1.50
	2'	0.33	0.21	0.31	0.12
	3'	0.26	0.31	0.26	0.26
	4'	0.67	0.68	0.53	0.49
	3	0.76	0.74	0.69	0.28
	4	0.83	0.88	0.67	0.45
	5	0.59	0.86	0.59	0.68
	6	0.20	0.33	0.27	0.20
	7	0.21	0.25	0.28	0.20
	8	0.65	0.71	0.64	0.76
	9	1.73	0.96	1.32	1.81
	10	1.20	1.09	0.95	1.01
1,2-Benzopyrene, **9**	1	1.33	0.81	1.01	1.49
	2	0.35	0.38	0.34	0.38
	3	1.35	0.86	1.07	1.54
	4	0.42	0.54	0.46	0.66
	5	0.78	0.91	0.71	0.80
	6	0.82	1.08	0.74	0.67
Perylene, **11**	1	1.48	0.55	0.94	0.84
	2	0.50	0.32	0.35	0.11
	3	0.83	0.65	0.56	0.30

[a]Calculated local anisotropy shift (reference 17).
[b]Uncoupled Hartree–Fock results from reference 34.
[c]Sum of the local anisotropic and delocalized contributions using ring-current calculations from references 4, 29, 45, and 46.

bond policy," in which the London approximation is used only for bonds at least four bonds distant from the proton considered, was employed. The calculated shifts are compared directly with the experimental shifts (as determined by Haigh and Mallion[1]) in the last column of Table 3.2, in contrast to the Haigh-Mallion procedure. Roberts claims better agreement with absolute chemical shifts than is observed with the McWeeny theory.[17]

Also appearing in Table 3.2 are calculations by Barfield, Grant, and Ikenberry,[27] based on their neighbor anisotropy chemical shift theory (see Chapter 2). Column I presents calculated values based on the neighbor anisotropy contribution alone, and Column III combines these values with a ring-current contribution based on an HMO ring-current model. Of particular interest and possible significance is the fact that the neighbor anisotropy calculation seems to handle the crowded protons as well as the uncrowded protons.

The theory of Vogler[28-30] includes both ring-current and neighbor anisotropy terms obtained using SCF techniques. Calculations based on this method are compared with experimental values in Table 3.3. Vogler's conclusion after detailed comparison of theory with experiment is that both contributions are important.

Table 3.3. Proton Chemical Shifts in Benzenoid Hydrocarbons

Compound	Proton	δ_{Exp}	$\delta_{Calc}{}^{a}$	$\delta_{Calc}{}^{b}$
Benzene, **1**	1	7.27	7.12	7.14
Naphthalene, **2**	1	7.73	7.70	7.65
	2	7.38	7.39	7.40
Anthracene, **3**	1	7.93	7.88	7.81
	2	7.39	7.45	7.47
	9	8.36	8.38	8.22
Phenanthrene, **4**	1	7.80	7.79	7.78
	2	7.51	7.48	7.50
	3	7.57	7.51	7.53
	4	8.62	8.58	8.59
	9	7.65	7.78	7.79
Chrysene, **5**	1	8.66	8.75	8.79
	2	7.93	7.98	7.97
	3	7.90	7.89	7.86
	4	7.52	7.53	7.55
	5	7.62	7.57	7.55
	6	8.72	8.70	8.72
Triphenylene, **6**	1	8.61	8.54	8.62
	2	7.58	7.51	7.56

Table 3.3. (Continued)

Compound	Proton	δ_{Exp}	$\delta_{Calc}{}^a$	$\delta_{Calc}{}^b$
Pyrene, **7**	1	8.00	8.04	7.99
	3	8.10	8.17	8.07
	4	7.93	7.86	7.84
1,2-Benzanthracene, **8**	1'	8.77	8.73	8.74
	2'	7.59	7.54	7.57
	3'	7.53	7.49	7.52
	4'	7.76	7.79	7.80
	3	7.55	7.76	7.80
	4	7.72	7.87	7.89
	5	7.95	7.94	7.88
	6	7.47	7.51	7.52
	7	7.47	7.51	7.53
	8	8.03	7.98	7.93
	9	9.08	9.19	9.11
	10	8.28	8.40	8.30
1,2-Benzopyrene	1	8.76	8.70	8.74
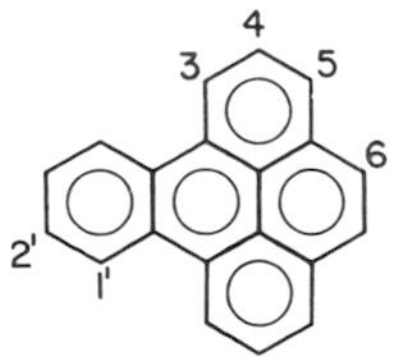	2	7.65	7.57	7.61
	3	8.81	8.89	8.31
	4	7.93	7.87	7.89
	5	8.07	8.16	8.10
	6	7.94	8.06	8.03

(9)

Compound	Proton	δ_{Exp}	$\delta_{Calc}{}^a$	$\delta_{Calc}{}^b$
Pentaphene	1	8.07	8.01	7.97
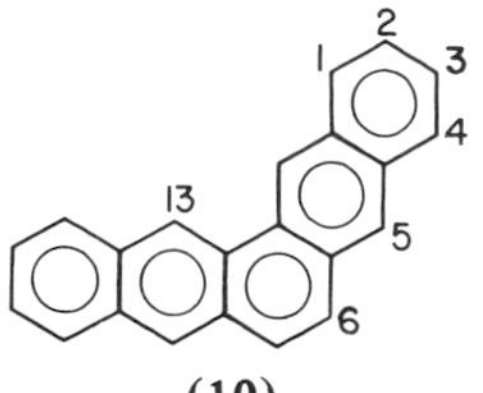	2	7.50	7.52	7.54
	3	7.48	7.51	7.53
	4	7.95	7.94	7.89
	5	8.18	8.36	8.30
	6	7.58	7.82	7.88
	13	9.18	9.29	9.23

(10)

Compound	Proton	δ_{Exp}	$\delta_{Calc}{}^a$	$\delta_{Calc}{}^b$
Perylene	1	8.11	8.34	8.49
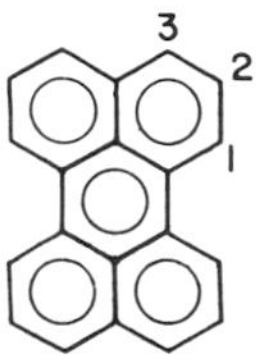	2	7.38	7.46	7.58
	3	7.57	7.72	7.75

(11)

Table 3.3. (Continued)

Compound	Proton	δ_{Exp}	$\delta_{Calc}{}^{a}$	$\delta_{Calc}{}^{b}$
1,2;3,4-Dibenzanthracene	1	8.68	8.65	8.74
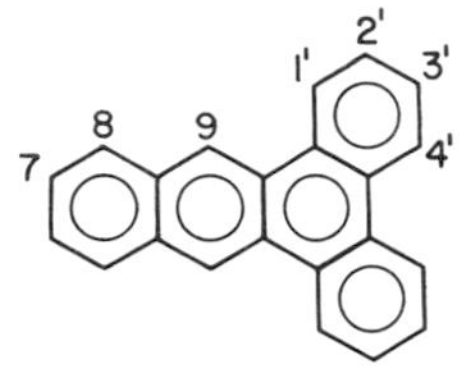	2	7.54	7.52	7.59
	3	7.53	7.50	7.57
	4	8.48	8.50	8.61
	7	7.46	7.52	7.54
	8	7.97	7.97	7.95
	9	8.97	9.09	9.10

(12)

Compound	Proton	δ_{Exp}	$\delta_{Calc}{}^{a}$	$\delta_{Calc}{}^{b}$
1,2;5,6-Dibenzanthracene	1	8.81	8.79	8.82
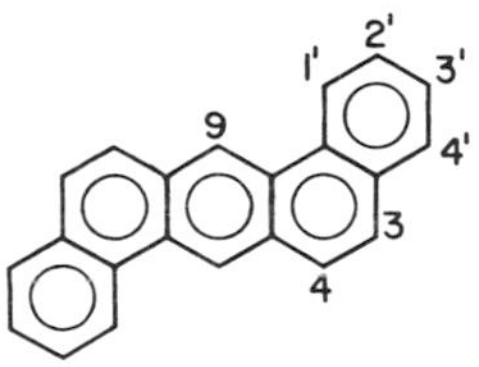	2	7.63	7.58	7.60
	3	7.55	7.52	7.55
	4	7.82	7.84	7.84
	3	7.67	7.85	7.88
	4	7.88	8.00	8.03
	9	9.08	9.25	9.22

(13)

Compound	Proton	δ_{Exp}	$\delta_{Calc}{}^{a}$	$\delta_{Calc}{}^{b}$
1,2;7,8-Dibenzanthracene	1	8.96	8.88	8.92
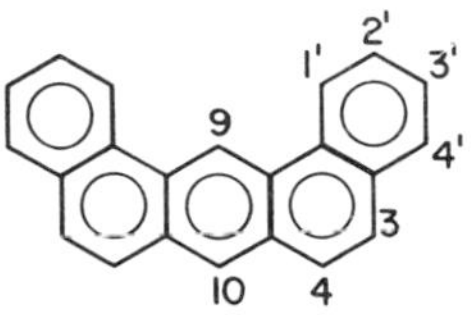	2	7.66	7.60	7.63
	3	7.56	7.53	7.56
	4	7.83	7.84	7.85
	3	7.66	7.84	7.87
	4	7.80	7.96	7.98
	9	9.98	9.61	9.63
	10	8.29	8.47	8.41

(14)

Compound	Proton	δ_{Exp}	$\delta_{Calc}{}^{a}$	$\delta_{Calc}{}^{b}$
Coronene 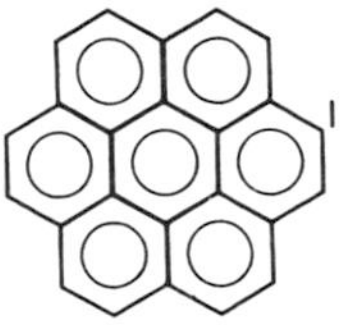	1	8.82	8.91	8.67

(15)

---

[a]Calculated by Vogler's method, including ring-current (RC) effect but not local anisotropy (LA) effect (reference 48).
[b]Calculated including both RC and LA effects (reference 48).

49

Annulenes

Proton chemical shifts in a number of annulenes[31-37] appear in Table 3.4. The observed chemical shift pattern in annulenes provided an early victory for a model attributing a significant contribution from delocalized ring currents as opposed to a model in which local currents alone are assumed to be important.[38] Consider Figure 3.3: both a model with local currents alone (I and II) and a ring-current model (III) predict that protons *outside* the ring experience a downfield shift—that is, the local field outside the ring caused by the induced current is in the same direction as the external applied field. *Inside* the ring, however, the models appear to disagree, the local current model predicting a downfield shift again, while the ring-current model predicts an upfield shift. The experimental observation of large upfield shifts of inner protons in (18)-annulene, **35**, and (14)-annulene, **24**, is cited by Gaidis and West[38] as evidence of the existence and significance of ring currents in these compounds.

Annulenes with $(4n + 2)$ mobile π electrons exhibit the characteristic pattern

Table 3.4. Proton Chemical Shifts for Several Annulenes

Compound	Proton	δ_H
1,6-Methano-[10]annulene	1	7.27
	2	6.95
	3	-0.52
1,6-Oxido-[10]annulene, **17**	1	7.46
	2	7.26
1,6-Imino-[10]annulene, **18**	1	7.41
	2	7.11
N-methyl-1,5-imino-[10]annulene, **19**	1	7.27
	2	7.00
11-Methylene-1,6-methano-[10]annulene	1	7.42
	2	7.02
	3	3.19
[12]Annulene	1	8.02
	2	5.65
	3,4	5.89

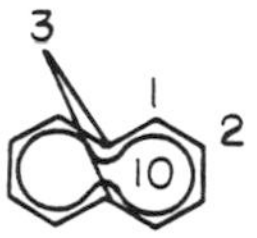

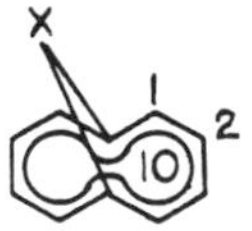

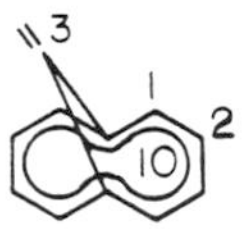

X = O (17)
X = NH (18)
X = NMc (19)

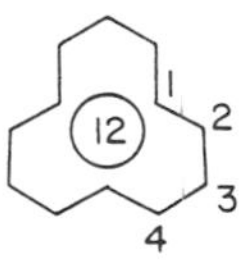

Table 3.4. (Continued)

Compound			Proton	δ_H
1,5,9-Tridehydro-[18]annulene 			1	4.45

(22)

Cyc[3,3,3]azine 			1	2.07
			2	3.65

(23)

[14]Annulene			1,2,4	7.60
			3	0.00

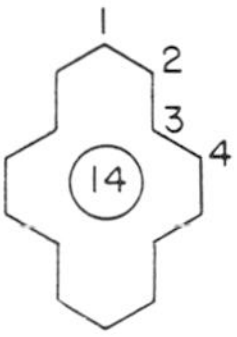

(24)

trans-15,16-Dimethyldihydropyrene			1	8.14
			2	8.63
			3	8.67
			4	-4.25

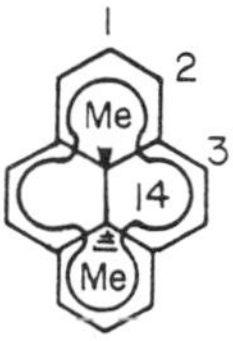

(25)

trans-15-16-Dihydropyrene 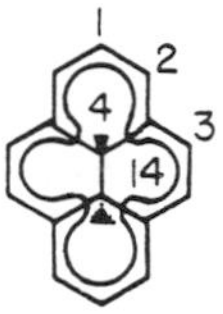			1	7.97
			2	8.50
			3	8.58
			4	-5.49

(26)

syn-1,6,8,13-Dioxido-[14]annulene, **27**			1,2	7.75
			3,4	7.60
			5	7.94

syn-1,6-Methano-8,13-oxido- [14]annulene, **28** 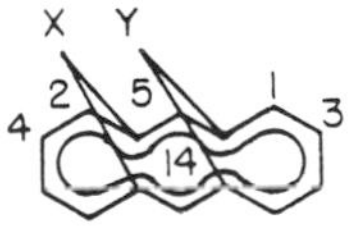	X = Y = O (27) X = O; Y = CH$_2$ (28)		1	7.65
			2	7.64
			3	7.34
			4	7.43
			5	7.75

Table 3.4. (Continued)

Compound	Proton	δ_H
1,6;8,13-Butano-[14]annulene	1	7.57
	2	7.12
	3	7.86

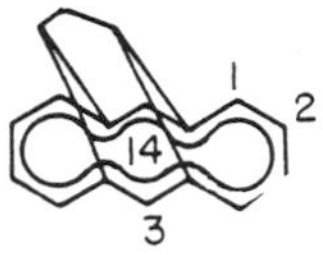

(29)

Compound	Proton	δ_H
1,6;8,13-Propano-[14]annulene	1	7.74
	2	7.55
	3	7.89
	4	-1.16
	5	-0.61

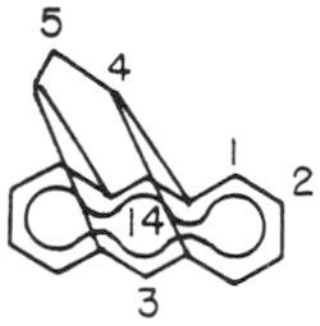

(30)

Compound	Proton	δ_H
1,8-Didehydro-[14]annulene	1	-5.48
	2	9.64
	3	8.54

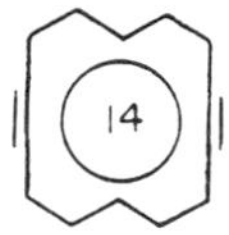

(31)

Compound	Proton	δ_H
[16]Annulene	1,2	5.20
	3	10.32

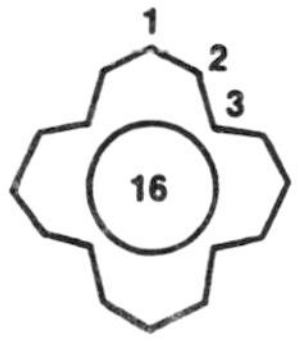

(32)

Compound	Proton	δ_H
trans-15,16-Dimethyldihydropyrenedianion	1	-3.30
	2	-3.40
	3	-3.96
	4	21.00

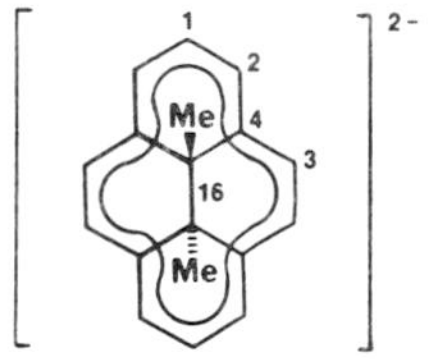

(33)

Compound	Proton	δ_H
[16]Annulene dianion	1	7.45
	2	8.83
	3	-8.17

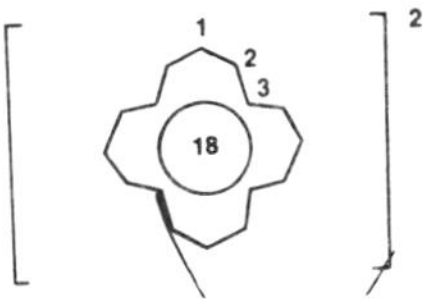

(34)

52

Table 3.4. (Continued)

Compound	Proton	δ_H
[18]Annulene	1	−2.26
	2	9.03

(35)

1,7,13-Tridehydro-[18]annulene	1	7.02
	2	8.10
	3	1.74
	4	7.56

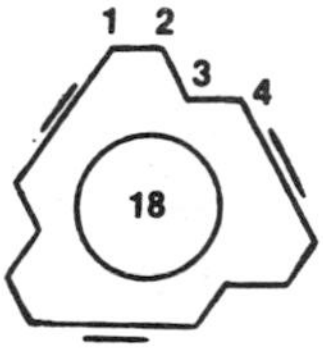

(36)

1,5,10,14-Tetramethyl-6,8,15,17- tetradehydro-[18]annulene	1	−5.24
	2	9.66

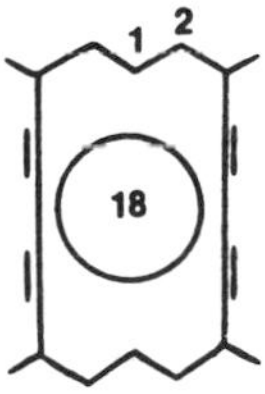

(37)

1,3,7,9,13,15-Hexadehydro- [18]annulene	1	7.02

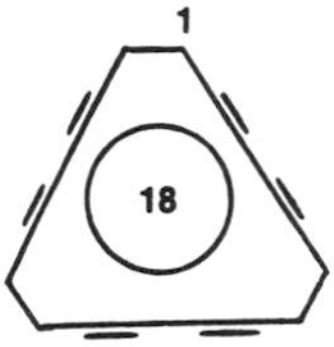

(38)

[24]Annulene	1,4	12.05
	2,3,5	4.73

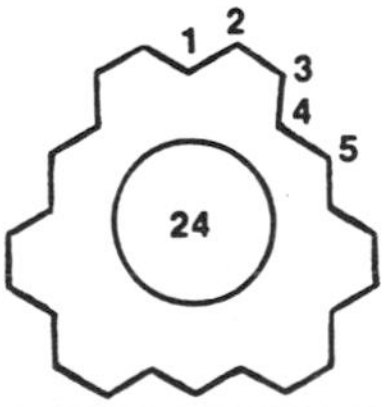

(39)

Source. References 31–37.

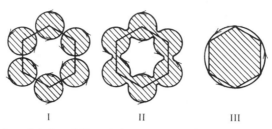

Figure 3.3. Local model (I and II) and ring-current model (III). Shaded areas indicate regions where protons experience upfield shift. (From reference 38.)

just described; annulenes with $4n$ mobile π electrons, such as bisdehydro(12)-annulene, however, exhibit the reverse pattern—upfield shifts for outer protons and downfield shifts for inner protons.[39] For this reason, $(4n + 2)$ annulenes are called *diatropic*, with "diamagnetic" ring currents, and the $(4n)$ annulenes are called *paratropic*, with "paramagnetic" ring currents. Both cases are adequately accounted for in ring-current theory: in the former case, the σ_1 term in Eq. 2.21 of McWeeny's theory dominates the σ_2 term; in the latter case, the reverse situation holds.[40]

Haddon's classical ring current model of the annulenes,[41] which contains few assumptions, accounts adequately for the observed chemical shift pattern for annulene protons.

There have been a number of quantum mechanical calculations of ring-current contributions to proton chemical shifts in annulenes.[33–35] Of particular interest are the comparisons of experimentally determined shifts with the predictions of Vogler's theory,[28–30] which incorporates both ring-current and neighbor anisotropy effects. Results for (14)-annulene, **24**, and (18)-annulene, **35**, were obtained. When geometric factors are taken into account, satisfactory agreement of theory with experiment is obtained.

Aromatic Nitrogen Heterocycles

There have been a large number of measurements of proton chemical shifts in aromatic nitrogen heterocyclic molecules.[42–53] An early calculation of chemical shifts for protons in such compounds was made by Veillard.[54] The theoretical method that he used to calculate ring-current contributions to the chemical shift was a slight modification of the method of McWeeny[17] (see Chapter 2). As was discussed in Chapter 2, the chemical shift of a particular nucleus is often considered to be the sum of terms representing local diamagnetic and paramagnetic currents on the atom in question, a contribution to local currents from other atoms, and a contribution from delocalized electrons moving in interatomic currents. Veillard's calculation was designed to test the hypothesis that for the protons in the heterocyclic molecules that he considered, all effects caused by local currents—that is, all effects other than those caused by ring currents—would be proportional to the charge density on the atom to which the proton is bound. Toward that end, he "corrected" the experimental chemical shift by adding a term designed to account theoretically for the ring-current effect and then plotted the corrected chemical shift against the charge density on the atom to which the particular proton was bound. He calculated the charge using SCF-MO theory. For a large number of compounds, including pyridine, **40**, pyrimidine, **42**, and related compounds, he obtained relatively satisfactory agreement of experiment with his hypothesis.

Hall, Hardisson, and Jackman[55-56] used coupled SCF theory (see Chapter 2) to calculate ring-current contributions to chemical shifts in benzenoid hydrocarbons and certain heterocyclic aromatic compounds. They concluded that the difference between HMO and SCF calculations is greater for heterocyclic compounds than for benzenoid hydrocarbons. It was also shown that ring-current contributions to chemical shifts in benzenoid hydrocarbons and the analogous aromatic nitrogen heterocyclics do not differ greatly. Using this observation, Cobb and Memory[57] again approached the hypothesis, described in Chapter 2, that local current contributions to the chemical shift would be proportional to the charge density on the atom to which the proton in question is bound.[58] Differences between the chemical shift of a proton in an aromatic nitrogen heterocyclic and the corresponding shift in the analogous benzenoid compound (Table 3.5) were plotted as a function of excess π-electron charge density on the

Table 3.5. Experimental Chemical Shifts between Protons in Nitrogen Heterocycles and the Corresponding Proton in the Analogous Benzenoid Compound

Compound		Proton	$\Delta\delta_H$
Pyridine		2	1.28
		3	-0.06
	(40)	4	0.35
Pyridazine		2	1.89
	(41)	3	0.19
Pyrimidine		2	1.91
		4	1.43
	(42)	5	0.01

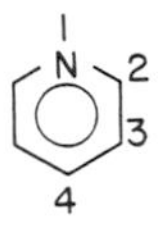

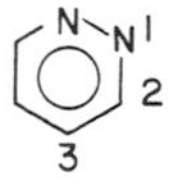

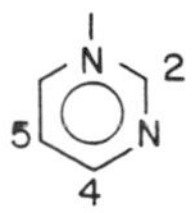

Table 3.5. (Continued)

Compound	Proton	$\Delta\delta_H$
Pyrazine	2	1.28
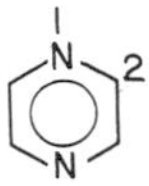 **(43)**		
Quinoline	2	1.44
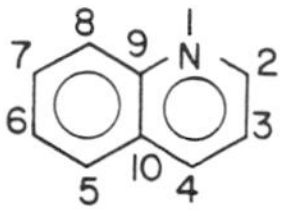	3	-0.11
	4	0.26
	5	-0.06
	6	0.06
(44)	7	0.24
	8	0.31
Isoquinoline	1	1.40
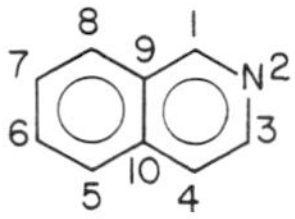	3	1.08
	4	-0.24
	5	-0.04
	6	0.20
(45)	7	0.12
	8	0.12
Cinnoline	3	1.85
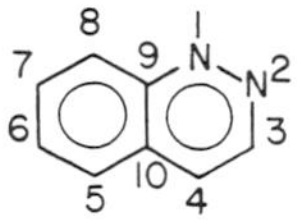	4	(0.02)
	5	(0.02)
	6	(0.39)
	7	(0.39)
(46)	8	0.74
Quinazoline	2	1.91
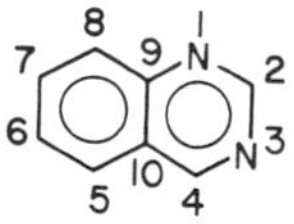	4	1.49
	5	0.10
	6	0.20
	7	0.46
(47)	8	0.26
Quinoxaline	2	1.36
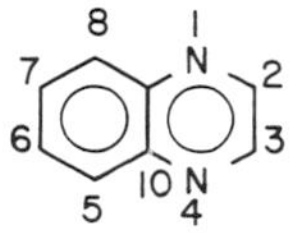	5	0.32
	6	0.30
(48)		
1,5-Naphthyridine	2	1.56
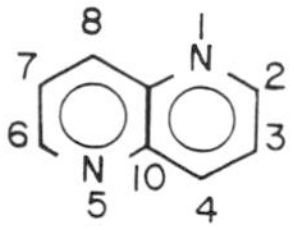	3	0.17
	4	0.61
(49)		

56

Table 3.5. **(Continued)**

Compound	Proton	$\Delta\delta_H$
1,6-Naphthyridine	2	1.69
	3	0.11
	4	0.49
	5	1.49
	7	1.35
(50)	8	0.14
Phthalazine	1	1.70
	5	0.19
(51)	6	0.48
2,6-Naphthyridine	1	1.60
	3	1.34
(52)	4	−0.01
Pteridine	2	2.00
	4	1.80
	6	1.73
(53)	7	1.50
Acridine	1	−0.10
	2	−0.00
	3	0.29
	4	0.30
(54)	10	0.22
Phenazine	1	0.28
(55)	2	0.42
5,6-Benzoquinoline	2	1.37
	3	−0.10
	4	0.22
	5	−0.06
	6	0.01
	7	0.09
	8	0.07
	9	0.20
(56)	10	0.32

Table 3.5. (Continued)

Compound	Proton	$\Delta\delta_H$
5,6-Benzoisoquinoline	1	1.41
	3	1.19
	4	−0.32
	5	−0.02
	6	(0.11)
	7	(0.15)
(57)	8	0.07
	9	0.14
	10	0.05
7,8-Benzoquinoline	1	0.28
	2	−0.07
	3	1.36
	5	0.68
	6	0.11
	7	0.11
(58)	8	−0.00
	9	0.09
	10	−0.07
3,4-Benzoisoquinoline	1	0.34
	2	0.28
	3	0.08
	4	−0.12
	5	−0.05
	6	0.16
(59)	7	0.13
	8	0.21
	9	1.53
1,5-Phenanthroline	1	0.29
	2	−0.06
	3	1.40
	5	0.85
	6	−0.03
	7	1.42
(60)	9	0.35
	10	0.12
1,8-Phenanthroline	2	1.41
	3	−0.08
	4	0.12
(61)	10	0.50

Table 3.5. (Continued)

Compound	Proton	$\Delta\delta_H$
4,5-Phenanthroline	1	0.33
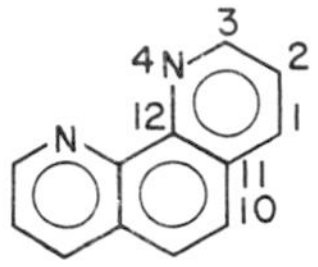	2	0.01
	3	1.55
	10	-0.00

(62)

3,4-Benzocinnoline	1	0.61
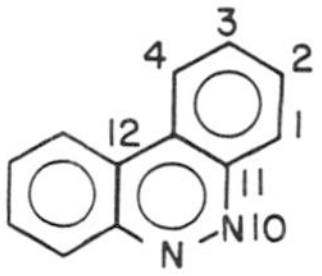	2	(0.25)
	3	(0.19)
	4	0.02

(63)

Naphtho(2,3-f)quinoline	1	(0.21)
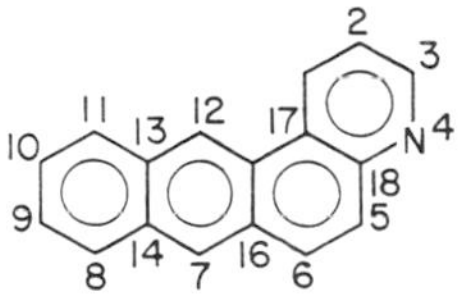	2	(-0.08)
	3	(1.39)
	5	(0.45)
	6	(0.33)
	7	0.09
	8	(0.08)
	9	(0.07)
	10	(0.06)
	11	(-0.00)
	12	(-0.08)

(64)

1,2-Benzacridine	1	(0.76)
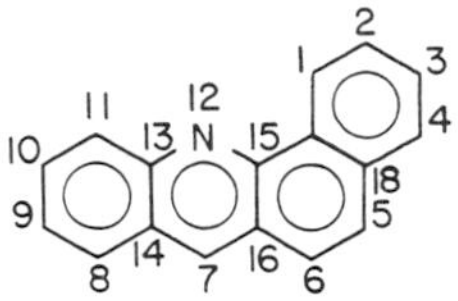	2	(0.05)
	3	(0.14)
	4	(-0.09)
	5	(0.12)
	6	(-0.00)
	7	(0.29)
	8	(-0.25)
	9	(0.20)
	10	(0.19)
	11	(0.34)

(65)

Naphtho(2,1-h)quinoline	2	1.56
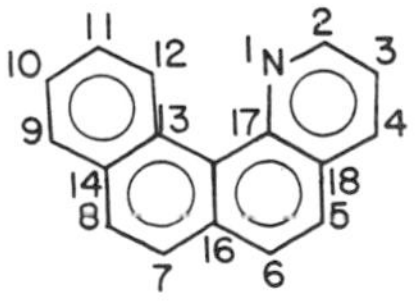	3	-0.12
	4	0.30
	5	
	6	0.04
	7	
	8	

(66)

Table 3.5. (Continued)

Compound	Proton	$\Delta\delta_H$
	9	-0.08
	10	-0.25
	11	
	12	-2.10
Naphtho(1,2-f)quinoline	1	(0.12)
	2	(0.25)
	3	(1.35)
	5	(0.33)
	6	(0.13)
	7	(-0.13)
(67)	8	(0.03)
	9	(-0.13)
	10	(-0.03)
	11	(-0.13)
	12	(-0.18)
Quino(5',6';5,6)quinoline	1	-0.10
	2	(-0.10)
	3	(1.40)
	5	0.32
(68)	6	0.17
Quino(8,7-f)quinoline	2	1.40
	3	-0.22
	4	0.14
	5	-0.00
	6	-0.17
(69)	7	0.07
	8	0.43
	10	1.33
	11	-0.05
	12	2.27
Naphtho(1',2';3,4)cinnoline	1	-0.25
	2	
	3	0.23
	4	(0.95)
	7	(0.70)
	8	(0.02)
(70)	9	(-0.10)
	10	
	11	0.23
	12	0.25

Table 3.5. (Continued)

Compound	Proton	$\Delta\delta_H$
Dibenzo(f,h)quinoline	2	1.30
	3	-0.15
	4	0.10
	5	(-0.12)
	6	(0.02)
	7	(0.02)
	8	(-0.12)
	9	(-0.12)
(71)	10	(0.02)
	11	(1.02)
	12	(0.72)
Thebenidine	1	(0.05)
	2	(0.28)
	3	(0.05)
	4	(0.20)
	5	(0.20)
	6	(0.05)
(72)	7	(0.28)
	8	(0.05)
	9	1.62
Naphtho(2,1-f)quinoline	2	(1.22)
	3	(0.15)
	4	(0.17)
	5	(-0.13)
	6	(0.05)
	7	(-0.17)
(73)	8	0.15
	9	
	10	0.17
	11	
	12	0.30

Source. References 42–53.

carbon to which the proton was bound. By hypothesis, the relationship should be linear. The result of this analysis is that the hypothesis is found to be satisfactory for protons bound to carbons in positions ortho to a nitrogen, so long as there is only one nitrogen atom in that particular ring (see Figure 3.4). Agreement of the original conjecture with experiment is not so good for other proton positions. Correlations were roughly the same for both HMO and SCF calculations of the charge densities.

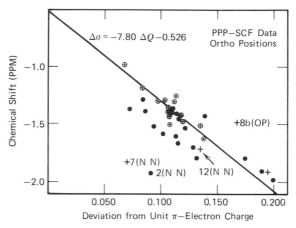

Figure 3.4. Charge density versus chemical shift for *ortho* positions; SCF data. Circled points are from molecules containing only a single nitrogen; others contain two or more. The divergent point 8b is from proton 4 of quinazoline (8). OP means this position is both *ortho* and *para*. N=N implies that there are directly bonded nitrogens in that molecule. Crosses are from data taken in CCl_4 solvents, dots from $CDCL_3$ solvents. (From reference 54.)

Miscellaneous Aromatic Compounds

The ring current contribution to chemical shifts in the nonalternant aromatic hydrocarbon azulene, **74**, has been calculated by Roberts.[59] The results are given in Table 3.6. It is clear that agreement between theory and experiment is not so satisfactory as in the case of hexagonal aromatic hydrocarbons. Indications are that inclusion of variations in π-electronic charge distribution and consideration of bond anisotropy effects do not substantially improve the correlation of theory with experiment. It is suggested that the sigma–pi separation

Table 3.6. δ_H in Azulene

Compounds	Proton	Theory	Experiment
Benzene, **1** (absolute value)		0.72	
Azulene	2	-0.05	0.54
	1,3	0.28	0.03
	4,8	0.67	0.96
	5,7	0.50	-0.22
(74)	6	0.49	0.22

Source. Reference 59.

inherent in all the calculations we have described may not be adequate to account for the experimental proton chemical shifts in nonalternant compounds.

The ring-current model has also provided an appropriate framework for consideration of chemical shifts in four-membered ring compounds[60] such as biphenylene, **75**, and related molecules (see Table 3.7).

Table 3.7. δ_H **of Compounds with a Four-Membered Ring**

Compounds	Proton	Chemical Shifts δ_H	
		Observed	Calculated
Biphenylene	1	6.60	6.34
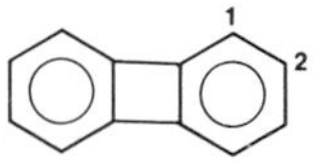	2	6.70	6.49
(75)			
Benzo[a]biphenylene	1		7.20
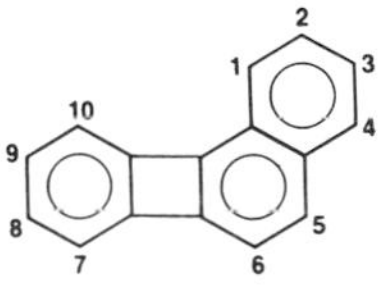	2	centered at	7.13
	3	7.25	7.14
	4		7.27
	5		7.02
	6	6.88	6.64
	7		6.09
(76)	8	centered at	6.27
	9	6.47	6.28
	10		6.14
Benzo[b]biphenylene	1	6.91	6.52
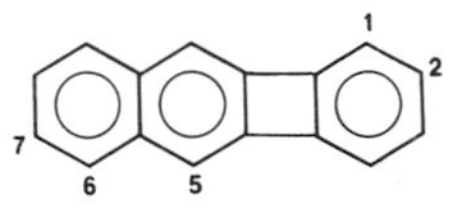	2	6.91	6.61
	5	6.91	6.76
	6	7.43	7.33
	7	7.23	7.23
(77)			
Benzo[a,c]biphenylene	1	centered at	7.41
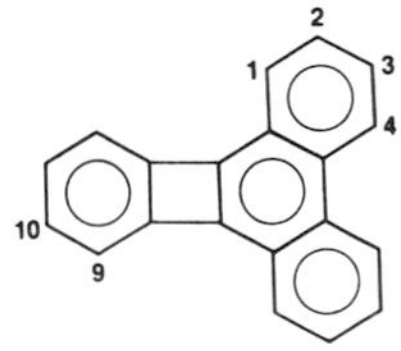	2	7.5	7.35
	3		7.38
	4	8.4	7.69
	9	centered at	6.00
	10	6.5	6.16
(78)			

Source. Reference 60.

Ege and Vogler[61] have considered the calculation of chemical shifts in several macrocyclic compounds, and Vogler[28-30] has applied his general theory (see Chapter 2) to an extensive number of compounds containing aromatic rings and a variety of double and triple bonds.

Ring-current theory has been used in a consideration of several five-membered heterocyclics, including thiophene, **79**, furan, and pyrrole.[62-65] Vincent and

(79)

others[67] have applied the McWeeny theory to a number of thiophenes and to thiazole. Again, support is obtained for the hypothesis that contributions to the chemical shift other than from the ring-current effect are proportional to the charge density on the carbon to which the proton is bonded. Mallion[68] has applied the McWeeny theory to some sulphur heterocyclic analogs of fluoranthene and to some large-ring aromatic nitrogen heterocyclics.

Out-of-Plane Protons

The agreement of experimental values of proton chemical shifts in aromatic compounds with predictions based on theories discussed in Chapter 2 differs, depending on whether the protons lie in or out of the plane of the aromatic rings. Rose,[69] for example, has observed that for protons not in the plane of an aromatic ring the chemical shifts predicted by McWeeny's theory[17] are not so close to experimental results as those predicted by the Johnson–Bovey semiclassical method.[70] In general, it seems that the Johnson–Bovey tables predict too high a deshielding for the region in and near the plane of an aromatic ring, while the Haigh–Mallion tables, which are based on the McWeeny theory and which give good results in the plane of the molecule, tend to underestimate the shielding in the region significantly out of the plane of the ring. Both the formula derived by Memory[71] and the predictions of the Roberts theory[18] give values intermediate between those of the Johnson–Bovey and Haigh–Mallion calculations for the chemical shift along the axis normal to the plane of the ring and in the plane of the ring.

Substituent Effects

A considerable amount of experimental work has been done to determine the effect of a substituent on the proton chemical shifts in a substituted benzene,

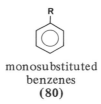

monosubstituted
benzenes
(80)

80.[72-78] Table 3.8 presents some of the experimental results for a set of mono-substituted benzenes, all with CCl_4 as a solvent and with extrapolation to infinite dilution. The results are presented as shifts from the δ_H for the benzene proton; positive values correspond to larger δ_H values—that is, toward higher frequency. These shifts may be used as guidelines for substituent effects in aromatic rings in

Table 3.8. Proton Chemical Shifts in Monosubstituted Benzenes, 80 (Shifts Relative to δ_H for the Benzene Proton)[a]

Substituent	$\Delta\delta_H$ (*Ortho*)	$\Delta\delta_H$ (*Meta*)	$\Delta\delta_H$ (*Para*)
$-CH_3$	-0.20	-0.12	-0.22
$-C(CH_3)_3$	0.02	-0.08	-0.21
$-NO_2$	0.95	0.26	0.38
$-COCl$	0.84	0.22	0.36
$-COBr$	0.80	0.21	0.37
$-SO_2Cl$	0.77	0.35	0.45
$-COOCH_3$	0.71	0.11	0.21
$-COOCH(CH_3)_2$	0.70	0.09	0.19
$-COOCH_2CH(CH_3)_2$	0.72	0.11	0.20
$-COCH_3$	0.62	0.14	0.21
$-COCH_2CH_3$	0.63	0.13	0.20
$-CHO$	0.56	0.22	0.29
$-SO_3CH_3$	0.60	0.26	0.33
$-CCl_3$	0.64	0.13	0.10
$-CN$	0.36	0.18	0.28
$-I$	0.39	-0.21	-0.00
$-Br$	0.18	-0.08	-0.04
$-Cl$	0.03	-0.02	-0.09
$-OCOCH_3$	-0.25	0.03	-0.13
$-OCH_3$	-0.48	-0.09	-0.44
$-OH$	-0.56	-0.12	-0.45
$-NH_2$	-0.75	-0.25	-0.65
$-NHCH_3$	-0.80	-0.22	-0.68
$-N(CH_3)_2$	-0.66	-0.18	-0.67

[a]See references 76–77.

general. They are also, to a certain extent, additive—that is, in disubstituted compounds, provided the substituents are not ortho to one another, the net effect on the chemical shift of a particular proton can be approximated by adding the effects of the substituents taken independently.[79]

It is reasonable to assume that the change in chemical shift caused by the substituent is related to a change in electron density arising from the effect of the substituent on the proton and the carbon to which it is bonded. Some suc-

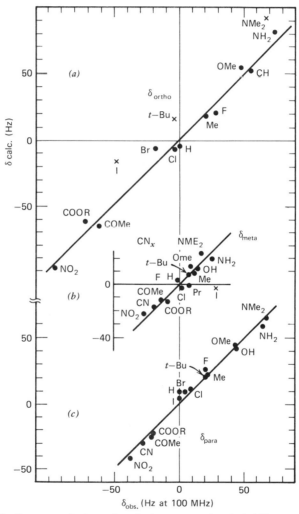

Figure 3.5. The linear correlations of each ring proton chemical shift with σ_i and σ_r constants. (*a*) *ortho*-, (*b*) *meta*-, (*c*) *para*-proton chemical shifts. (From reference 77.)

cess has been achieved in relating the change in chemical shift to the Hammett constant or Pauling electronegativity of the substituent, particularly for para-protons.[79,80] Better correlation with the experimental data is obtained using the dual-substituent parameter (DSP) method.[81,82] In this method the following equation is used:

$$\delta = \sigma_i \rho_i + \sigma_r \rho_r \tag{3.2}$$

where σ_i is a measure of the inductive effect of the substituent and is related to its polarity, and σ_r is a measure of the resonance effect of the substituent. The mixing parameters ρ_i and ρ_r depend on the position of the substituent with respect to the proton whose chemical shielding parameter is to be determined. A plot relating experimental chemical shifts to calculated values based on the DSP method is given in Figure 3.5. (Note that the figure, which is from reference 77, has the sign of the shift opposite to that of the convention used in this book.)

More recently, an analysis of substituent effects on proton chemical shifts has involved the application of the DSP method to meta-substituted nitrobenzenes.[83]

Proton Chemical Shifts in Aromatic Ions

Proton chemical shifts have been determined for a number of aromatic cations and anions.[84–104] In general, the shifts in cations are to even greater δ_H values than for the neutral counterparts. The reason is that the deficiency of electron density has a deshielding effect operating in the same direction as the ring-current shift. Conversely, the extra negative charge in the anions produces a diamagnetic shift opposing the ring-current effect, giving δ_H values smaller than for the neutral counterparts.

Summary

In this section, a few rules are given to help in estimating the chemical shift for a proton in an aromatic compound. The estimates will be rough, but may be useful to the practicing chemist in making order-of-magnitude estimates.

1. For a proton bonded to an aromatic ring, start with an estimate of δ_H 7.3, approximately the proton chemical shift in benzene.
2. To estimate the ring-current effect of other aromatic rings in a molecule, use Figure 2.2 or 2.3. If the chemical shift is to be expressed on the δ scale, note that the signs on Figures 2.2 and 2.3 must be reversed; the effect of currents in other rings in the same plane will be to *increase* the delta value. The effect of other rings may also be estimated by using the

fact that the ring-current chemical shift in benzene is about 1.5 ppm, and the fact that the effect falls off roughly as the cube of the distance from the ring center. Therefore one can estimate the effect of a distance by the equation

$$\Delta\delta_H = 1.5\left(\frac{r}{R}\right)^3 \qquad (3.3)$$

where R is the distance of the proton from the distant ring center and r is the distance from the adjacent ring center to the proton.

3. To estimate the effect of a substituent, one may use Table 3.8. The substituent effect differs, of course, depending on the distance of the proton from the substituent—that is, whether the proton is ortho, meta, or para to the substitutent. A rule of thumb is that the effects of more than one substituent are additive.

4. A proton ortho to a nitrogen in an aromatic ring has its δ_H shifted to a higher value, about 8.5.

^{19}F CHEMICAL SHIFTS

Considerable experimental work has been done on fluorine NMR in aromatic compounds.[105-120] Karplus and Das[105] applied Eq. 2.63 to calculate ^{19}F chemical shifts in a variety of multifluorobenzenes; the constant σ_0 was calculated using the molecules F_2 and monofluorobenzene as references. Dewar and Kelemen[106] considered the ^{19}F chemical shifts in a large number of aromatic compounds (see Table 3.9). They found that the Karplus–Das theory, which implies that the dominant factor in the chemical shift of a fluorine nucleus is the

Table 3.9. ^{19}F Chemical Shifts in Aromatic Compounds

Compound	δ_F	Solvent	Reference
Fluorobenzene	113.2	*a*	107
1,2-Difluorobenzene	139.1	*b*	106
1,3-Difluorobenzene	110.1	*b*	106
1,4-Difluorobenzene	119.6	*b*	106
1̲,2,4-Trifluorobenzene	143.6	*b*	111
1,2̲,4-Trifluorobenzene	134.1	*b*	111
1,2,4̲-Trifluorobenzene	115.6	*b*	111
1̲,3,5̲-Trifluorobenzene	107.7	*b*	111
1̲,2,3,4-Tetrafluorobenzene	139.9	*c*	108
1,2̲,3,4-Tetrafluorobenzene	157.4	*a*	109
1̲,2,3,5-Tetrafluorobenzene	132.5	*a*	108
1,2̲,3,5-Tetrafluorobenzene	167.0	*c*	109
1,2,3,5̲-Tetrafluorobenzene	113.0	*c*	108

Table 3.9. (Continued)

Compound	δ_F	Solvent	Reference
1,2,4,5-Tetrafluorobenzene	139.9	b	111
1,2,3,4,5-Pentafluorobenzene	138.9	d	116
1,2,3,4,5-Pentafluorobenzene	162.1	d	116
1,2,3,4,5-Pentafluorobenzene	153.5	b	116
Hexafluorobenzene	162.3	d	116

(Underlined position numbers: 1,2,4,5-Tetrafluorobenzene; 1,2,3,4,5-Pentafluorobenzene; 1,2,3,4,5-Pentafluorobenzene; 1,2,3,4,5-Pentafluorobenzene)

(81)

Compound	δ_F	Solvent	Reference
2-Fluorostyrene	119.1	c	108
3-Fluorostyrene	112.8	d	109
4-Fluorostyrene	114.6	d	109

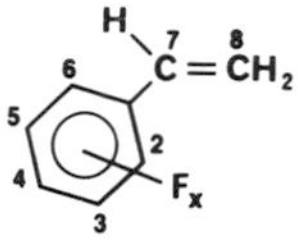

(82)

Compound	δ_F	Solvent	Reference
2,3,4,5,6-Pentafluorostyrene	144.3	d	118
2,3,4,5,6-Pentafluorostyrene	163.8	d	118
2,3,4,5,6-Pentafluorostyrene	156.8	d	118

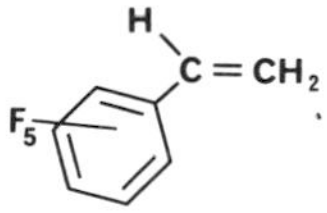

(83)

Compound	δ_F	Solvent	Reference
2,3,4,5,6,7,8,8-Octafluorostyrene	161.6	d	118
2,3,4,5,6,7,8,8-Octafluorostyrene	149.7	d	118

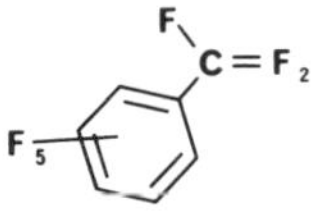

(84)

Compound	δ_F	Solvent	Reference
3-Fluorobiphenyl	113.1	a	112

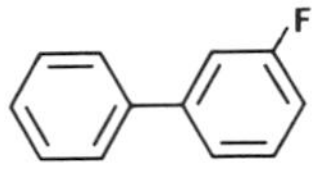

(85)

Compound	δ_F	Solvent	Reference
4-Fluorobiphenyl	115.9	a	112

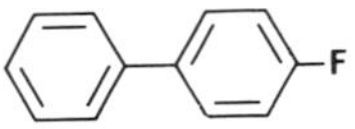

(86)

Compound	δ_F	Solvent	Reference
3,3'-Difluorobiphenyl	112.8	c	108

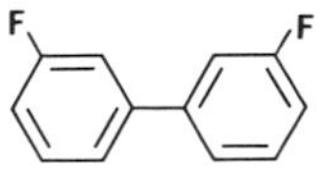

(87)

Compound	δ_F	Solvent	Reference
4,4'-Difluorobiphenyl	115.8	c	108

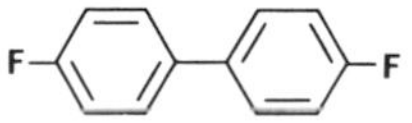

(88)

69

Table 3.9. (Continued)

Compound	δ_F	Solvent	Reference
1-Fluoronaphthalene	123.5	*a*	108

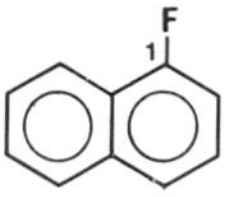

(89)

2-Fluoronaphthalene	114.9	*a*	112

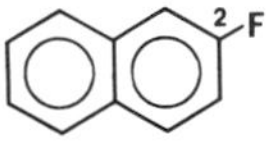

(90)

1-Fluoroacenaphthylene	134.4	*c*	108

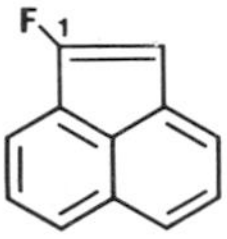

(91)

3-Fluoroacenaphthylene	110.6	*c*	108
4-Fluoroacenaphthylene	113.3	*c*	108
5-Fluoroacenaphthylene	122.8	*c*	108

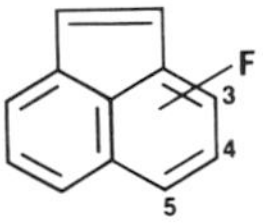

(92)

1-Fluoroanthracene	122.7	*c*	108

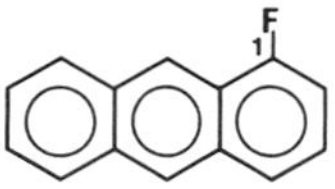

(93)

2-Fluoroanthracene	114.1	*c*	108

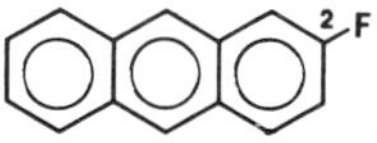

(94)

9-Fluoroanthracene	130.9	*c*	112

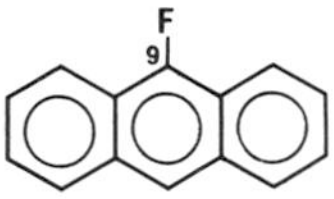

(95)

1-Fluorophenanthrene	122.4	*a*	112
2-Fluorophenanthrene	115.3	*a*	112
3-Fluorophenanthrene	113.5	*a*	112
4-Fluorophenanthrene	109.9	*a*	112

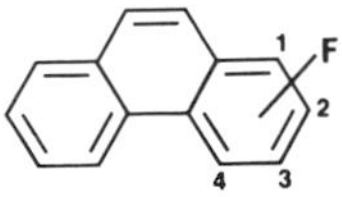

(96)

Table 3.9. (Continued)

Compound	δ_F	Solvent	Reference
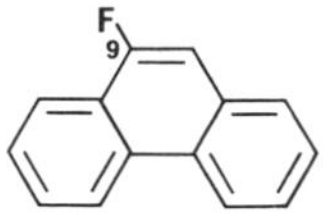 9-Fluorophenanthrene	125.3	a	112

(97)

| 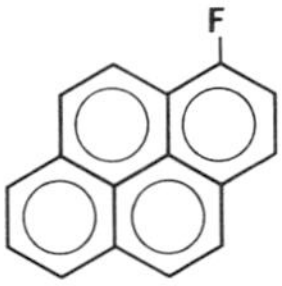 1-Fluoropyrene | 123.3 | a | 112 |

(98)

| 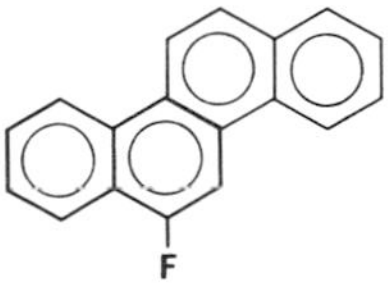 6-Fluorochrysene | 123.4 | a | 112 |

(99)

| 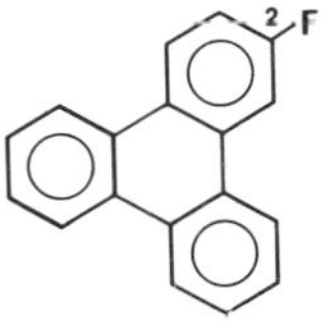 2-Fluorotriphenylene | 114.2 | a | 109 |

(100)

| 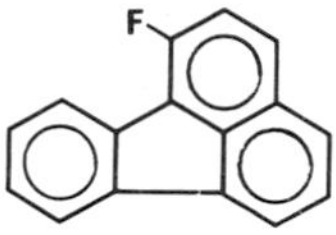 1-Fluorofluoranthene | 113.9 | c | 108 |

(101)

| 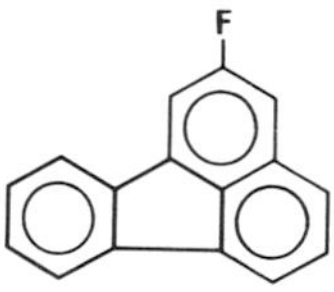 2-Fluorofluoranthene | 112.1 | c | 112 |

(102)

| 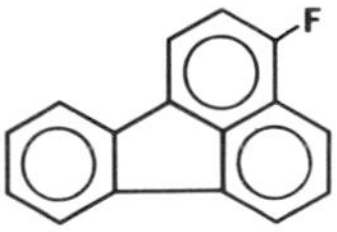 3-Fluorofluoranthene | 122.9 | a | 108 |

(103)

Table 3.9. (Continued)

Compound	δ_F	Solvent	Reference
8-Fluorofluoranthene	114.3	c	108

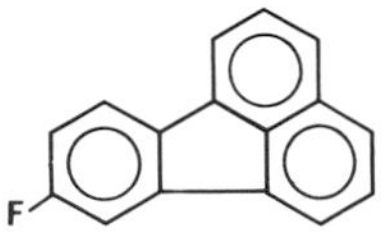

(104)

Compound	δ_F	Solvent	Reference
2-Fluoropyridine	67.9	c	108
3-Fluoropyridine	126.8	c	114
2,3,4,5,6-Pentafluoropyridine	87.6	e	114
2,3,4,5,6-Pentafluoropyridine	162.0	e	114
2,3,4,5,6-Pentafluoropyridine	134.1	e	115

(105)

Compound	δ_F	Solvent	Reference
4-Phenyl-2,3,5,6-tetrafluoropyridine	92.8	f	113

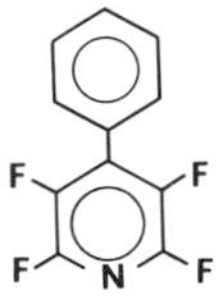

(106)

Compound	δ_F	Solvent	Reference
2,4,6-Trifluoropyrimidine	43.0	b	113
2,4,6-Trifluoropyrimidine	55.7	b	113
2,4,5,6-Tetrafluoropyrimidine	47.3	b	117

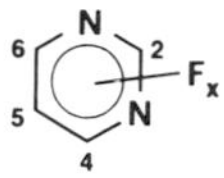

(107)

Compound	δ_F	Solvent	Reference
3,4,5,6-Tetrafluoropyridazine	82.7	b	117
3,4,5,6-Tetrafluoropyridazine	144.3	d	117

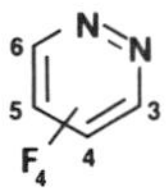

(108)

Compound	δ_F	Solvent	Reference
Tetrafluoropyrazine	93.9	b	112

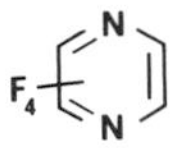

(109)

Compound	δ_F	Solvent	Reference
2-Fluoroquinoline	62.7	a	112
3-Fluoroquinoline	129.3	a	112
5-Fluoroquinoline	123.9	a	112

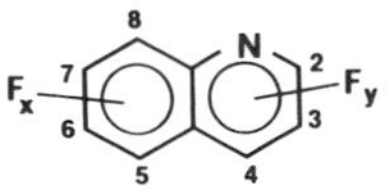

(110)

Table 3.9. (Continued)

Compound	δ_F	Solvent	Reference
6-Fluoroquinoline	114.2	*a*	112
7-Fluoroquinoline	110.4	*a*	112
8-Fluoroquinoline	126.5	*a*	114
3,4,5,6,7,8-Hexafluoroquinoline	133.3	*g*	114
3,4,5,6,7,8-Hexafluoroquinoline	146.0	*g*	114
2,3,4,5,6,7,8-Heptafluoroquinoline	77.2	*g*	114
2,3,4,5,6,7,8-Heptafluoroquinoline	133.3	*g*	114
2,3,4,5,6,7,8-Heptafluoroquinoline	145.7	*h*	114

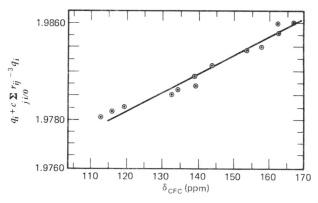

(111)

5-Fluoroisoquinoline	123.9	*c*	109
3,4,5,6,7,8-Hexafluoroisoquinoline	97.2	*e*	114
1,3,4,5,6,7,8-Heptafluoroisoquinoline	61.0	*e*	114
1,3,4,5,6,7,8-Heptafluoroisoquinoline	96.5	*e*	114

Source. References 106–118.

Solvents: *a*, in chloroform; *b*, neat; *c*, in dimethylformamide; *d*, in carbon tetrachloride; *e*, in acetone; *f*, approximate value in fluorotrichloromethane; *g*, in dichloromethane; *h*, see reference 117.

π-electron density near the fluorine atom, gives satisfactory agreement with experiments as long as both positions ortho to the fluorine are occupied by hydrogen atoms; in cases in which there are one or two fluorines ortho to the fluorine in question, agreement fails. Dewar and Kelemen[106] showed that this discrepancy could not be attributed to sigma polarization. When, however, they extended the Karplus-Das-Prosser-Goodman theory to include range effects as given in Eq. 2.64, quite satisfactory agreement with experiments was shown (Figure 3.6).

Figure 3.6. Comparison of the Dewar-Kelemen theory with experiments on ^{19}F chemical shifts in fluorobenzens. (From reference 106.)

When the long-range interactions are taken into account, the "ortho effect" disappears. The charge densities used in the calculation were obtained using an SCF method.

An encyclopedic listing of fluorine chemical shifts in aromatic compounds is given in reference 120.

REFERENCES

1. C. W. Haigh and R. B. Mallion, *Mol. Phys.* **18**, 737 (1970).
2. H. J. Bernstein, W. G. Schneider, and J. A. Pople, *Proc. R. Soc.* **A236**, 515 (1956).
3. G. G. Hall and A. Hardisson, *Proc. R. Soc.* **A268**, 328 (1962).
4. N. Jonathan, S. Gordon, and B. P. Dailey, *J. Chem. Phys.* **36**, 2443 (1962).
5. J. D. Memory, G. W. Parker, and J. C. Halsey, *J. Chem. Phys.* **45**, 3567 (1966).
6. J. D. Memory and T. B. Cobb, *J. Chem. Phys.* **47**, 2020 (1967).
7. H. P. Figeys, *Tetrahedron Lett.* **38**, 4625 (1966).
8. R. C. Fahey and G. C. Graham, *J. Chem. Phys.* **69**, 4417 (1968).
9. R. H. Martin, N. Defay, F. Geerts-Evrard, and S. Delavarenne, *Tetrahedron* **20**, 1073 (1964).
10. T. J. Batterham, L. Tsai, and H. Ziffer, *Aust. J. Chem.* **17**, 163 (1964).
11. T. J. Batterham, L. Tsai, and H. Ziffer, *Aust. J. Chem.* **18**, 1959 (1965).
12. C. W. Haigh and R. B. Mallion, *J. Mol. Spectrosc.* **29**, 478 (1969).
13. K. D. Bartle and J. A. S. Smith, *Spectrochim. Acta* **A23**, 1689 (1967).
14. M. A. Cooper and S. L. Manatt, *J. Am. Chem. Soc.* **91**, 6325 (1969).
15. L. Cassidei and O. Sciacovelli, *Org. Magn. Resonance* **15**, 257 (1981).
16. C. W. Haigh, R. B. Mallion, and E. A. G. Armour, *Mol. Phys.* **18**, 751 (1970).
17. R. McWeeny, *Mol. Phys.* **1**, 311 (1958).
18. H. G. Ff. Roberts, *Mol. Phys.* **27**, 843 (1974).
19. H. Spiesecke and W. G. Schneider, *J. Chem. Phys.* **35**, 731 (1961).
20. J. S. Waugh and R. W. Fessenden, *J. Am. Chem. Soc.* **79**, 846 (1957).
21. R. B. Mallion, *J. Chem. Phys.* **75**, 793 (1981).
22. P. H. Blustin, *Chem. Phys. Lett.* **64**, 507 (1979).
23. P. H. Blustin, *Mol. Phys.* **39**, 565 (1980).
24. C. Reid, *J. Mol. Spectrosc.* **1**, 18 (1957).
25. K. D. Bartle and J. A. S. Smith, *Spectrochim. Acta* **A23**, 1689 (1967).
26. B. V. Cheney, *J. Am. Chem. Soc.* **90**, 5386 (1968).
27. M. Barfield, D. M. Grant, and D. Ikenberry, *J. Am. Chem. Soc.* **97**, 6956 (1975).
28. H. Vogler, *J. Am. Chem. Soc.* **100**, 7464 (1978).
29. H. Vogler, *J. Mol. Struc.* **51**, 289 (1979)
30. H. Vogler, *Tetrahedron* **36**, 657 (1979).
31. F. A. Bovey, *NMR Data Tables for Organic Compounds*, Vol. 1, Interscience, New York, 1967.
32. H. Gunther, *Z. Naturforsch.* **B20**, 948 (1965).

33. H. Gunther and H.-H. Hinrichs, *Tetrahedron* **24**, 7033 (1968).
34. J. F. M. Oth, J.-M. Gilles, and G. Schroder, *Tetrahedron Lett.* **67** (1970).
35. K. G. Untch and D. C. Wysocki, *J. Am. Chem. Soc.* **88**, 2608 (1966).
36. D. Farquhar and D. Leaver, *Chem. Commun.* **1969**, 24 (1969).
37. Y. Gaoni and F. Sondheimer, *Proc. Chem. Soc.* **1964**, 299 (1964).
38. J. M. Gaidis and R. West, *J. Chem. Phys.* **46**, 1218 (1967).
39. F. Sondheimer, *Acc. Chem. Res.* **5**, 81 (1972).
40. J. A. Pople and K. G. Untch, *J. Am. Chem. Soc.* **88**, 4811 (1966).
41. R. C. Haddon, *Tetrahedron* **28**, 3613 (1972).
42. P. J. Black, R. D. Brown, and M. L. Heffernan, *Aust. J. Chem.* **20**, 1305 (1967); **17**, 558 (1964); **19**, 1287 (1966).
43. P. J. Black, R. D. Brown, and M. L. Heffernan, *Aust. J. Chem.* **20**, 1325 (1967).
44. R. J. Chuck and E. W. Randall, *J. Chem. Soc.* **B1967**, 261 (1967).
45. K. Tori and M. Ogata, *Chem. Pharm. Bull. (Tokyo)* **12**, 272 (1964).
46. G. S. Reddy, R. T. Hobgood, and J. H. Goldstein, *J. Am. Chem. Soc.* **84**, 336 (1962).
47. W. W. Paudler and T. J. Kress, *J. Heterocycl. Chem.* **2**, 393 (1965).
48. N. P. Buu-Hoi, M. Dufour, and P. Jacquignon, *Académie des Sciences, Paris, Comptes Rendus* **C263**, 1448 (1966).
49. J. P. Kokko and J. H. Goldstein, *Spectrochim. Acta* **19**, 1119 (1963).
50. D. J. Blears and S. S. Dangluk, *Tetrahedron* **23**, 2927 (1967).
51. B. P. Dailey, A. Gawer, and W. C. Heikam, *Discuss. Faraday Soc.* **34**, 18 (1962).
52. A. H. Gawer and B. P. Dailey, *J. Chem. Phys.* **42**, 2658 (1960).
53. R. H. Martin, N. Defay, F. Geerts-Evrard, and D. Bogaert-Verhoogen, *Tetrahedron* **1966**, Supp. 8, 181 (1966).
54. A. Veillard, *J. Chim. Phys.* **59**, 1056 (1962).
55. G. G. Hall and A. Hardisson, *Proc. R. Soc.* **A268**, 328 (1962).
56. G. G. Hall, A. Hardisson, and L. M. Jackman, *Tetrahedron* **19** Supp. **2**, 101 (1963).
57. T. B. Cobb and J. D. Memory, *J. Chem. Phys.* **50**, 4265 (1969).
58. G. Fraenkel, R. E. Carter, A. McLachlan, and J. H. Richards, *J. Am. Chem. Soc.* **82**, 5846 (1960).
59. H. G. Ff. Roberts, *J. Magn. Resonance* **29**, 7 (1978).
60. H. P. Figeys, N. Defay, R. H. Martin, J. F. W. McOmie, B. E. Ayres, and J. B. Chadwick, *Tetrahedron* **32**, 2571 (1976).
61. G. Ege and H. Vogler, *Z. Naturforsch.* **27b**, 1164 (1972); *Tetrahedron* **31**, 569 (1975); *Tetrahedron* **32**, 1789 (1976).
62. E. Corradi, P. L. Lazzeretti, and F. Taddei, *Mol. Phys.* **26**, 41 (1973).
63. D. F. Ewing and R. M. Scrowston, *Org. Magn. Resonance* **3**, 405 (1971).
64. W. Brugel, *Nuclear Magnetic Resonance Spectra and Chemical Structure*, Vol. 1, Academic, New York, 1967.
65. B. Clin and B. Lemanceau, *J. Chem. Phys.* **66**, 1327 (1966).
66. L. M. Jackman and S. Sternhell, *Applications of Nuclear Magnetic Resonance Spectroscopy in Organic Chemistry*, 2nd ed., p. 266, Pergamon, New York, 1969.
67. E. J. Vincent, R. Phan-Tan-Luu, and J. Metzger, *Bull. Soc. Chim. Fr.* **1966**, 3537 (1966).

68. R. B. Mallion, *J. Chem. Soc. Perkin Trans. II* **1973**, 235 (1973); *Biochim.* **56**, 197 (1974).

69. P. I. Rose, *Org. Magn. Resonance* **5**, 187 (1973).

70. C. E. Johnson, Jr. and F. A. Bovey, *J. Chem. Phys.* **29**, 1012 (1958).

71. J. D. Memory, *J. Magn. Resonance* **27**, 241 (1977).

72. J. M. Read and J. H. Goldstein, *J. Mol. Spectrosc.* **23**, 179 (1967).

73. H. B. Evans, A. R. Tarpley, and J. H. Goldstein, *J. Phys. Chem.* **72**, 2552 (1968).

74. S. Castellano, C. Sun, and R. Kostelnik, *Tetrahedron Lett.* **51**, 5205 (1967).

75. K. Hayamizu and O. Yamamoto, *J. Mol. Spectrosc.* **25**, 422 (1968).

76. K. Hayamizu and O. Yamamoto, *J. Mol. Spectrosc.* **28**, 89 (1968).

77. K. Hayamizu and O. Yamamoto, *J. Mol. Spectrosc.* **29**, 183 (1969).

78. Y. Yukawa, Y. Tsuno, and N. Shimizu, *Bull. Chem. Soc. Japan* **44**, 2843 (1971).

79. L. M. Jackman and S. Sternhell, *Applications of Nuclear Magnetic Resonance Spectroscopy*, 2nd ed., Chapters 3–6, Pergamon, New York, 1969.

80. M. T. Tribble and J. G. Traynham, *in* N. B. Chapman and J. Shorter (Eds.): *Advances in Linear Free Energy Relationships*, p. 143, Plenum, London, 1972.

81. W. J. Hehre, R. W. Taft, and R. D. Topsom, *Prog. Phys. Org. Chem.* **12**, 159 (1976).

82. S. Ehrenson, R. T. C. Brownlee, and R. W. Taft, *Prog. Phys. Org. Chem.* **10**, 1 (1973).

83. B. D. Batts and G. Pallos, *Org. Magn. Resonance* **13**, 349 (1980).

84. T. J. Katz, *J. Am. Chem. Soc.* **82**, 3785 (1960).

85. H. P. Fritz, and H. Keller, *Z. Naturforsch.* **16b**, 231 (1961).

86. G. Fraenkel, R. E. Carter, A. McLachlan, and J. H. Richards, *J. Am. Chem. Soc.* **82**, 5846 (1960).

87. S. McLean and P. Haynes, *Can. J. Chem.* **41**, 1231 (1963).

88. T. Schaefer and W. G. Schneider, *Can. J. Chem.* **41**, 966 (1963).

89. T. J. Katz, M. Yoshida, and L. C. Siew, *J. Am. Chem. Soc.* **87**, 4516 (1965).

90. T. J. Katz, M. Rosenberger, and R. K. O'Hara, *J. Am. Chem. Soc.* **86**, 249 (1964).

91. T. J. Katz, *J. Am. Chem. Soc.* **82**, 3785 (1960).

92. H. P. Fritz and H. Keller, *Z. Naturforsch.* **16b**, 231 (1961).

93. T. J. Katz and P. J. Garratt, *J. Am. Chem. Soc.* **86**, 5194 (1964).

94. E. A. LaLancette and R. E. Benson, *J. Am. Chem. Soc.* **87**, 1941 (1965).

95. M. Ogliaruso, R. Rieke, and S. Winstein, *J. Am. Chem. Soc.* **88**, 4731 (1969).

96. P. Radlick and W. Rosen, *J. Am. Chem. Soc.* **88**, 3461 (1966).

97. G. Fraenkel, R. E. Carter, A. McLachlan, and J. H. Richards, *J. Am. Chem. Soc.* **82**, 5846 (1960).

98. J. E. Mahler, D. A. K. Jones, and R. Pettit, *J. Am. Chem. Soc.* **86**, 3589 (1964).

99. R. Breslow, H. Hover, and H. W. Change, *J. Am. Chem. Soc.* **84**, 3168 (1962).

100. J. L. Rosenberg, Jr., J. E. Mahler, and R. Pettit, *J. Am. Chem. Soc.* **84**, 2842 (1962).

101. C. E. Keller and R. Pettit, *J. Am. Chem. Soc.* **88**, 604 (1966).

102. R. J. Smith and R. M. Pagni, *J. Am. Chem. Soc.* **101**, 4769 (1979).

103. F. Van de Griendt and H. Cerfontain, *Tetrahedron* **35**, 2563 (1979).

104. F. Van de Griendt and H. Cerfontain, *Tetrahedron* **36**, 317 (1980).

105. M. Karplus and T. P. Das, *J. Chem. Phys.* **34**, 1683 (1961).

106. M. J. S. Dewar and J. Kelemen, *J. Chem. Phys.* **49**, 499 (1968).

107. M. J. S. Dewar, R. C. Fahey, and P. J. Grisdale, *Tetrahedron Lett.* **1963**, 343 (1963).

108. M. J. S. Dewar and J. Michl (unpublished; see reference 106).

109. R. W. Taft, E. Price, J. R. Fox, J. C. Lewis, K. K. Andersen, and G. T. Davis, *J. Am. Chem. Soc.* **85**, 3146 (1963).

110. M. J. S. Dewar and C. de Llano (unpublished; see reference 106).

111. H. S. Gutowsky, D. W. McCall, B. R. McGarvey, and L. H. Meyer, *J. Am. Chem. Soc.* **74**, 4809 (1952).

112. M. J. S. Dewar and R. C. Fahey (unpublished; see reference 106).

113. H. Schroeder, E. Kober, H. Ulrich, R. Ratz, H. Agahigian, and C. Grundmann, *J. Org. Chem.* **27**, 2580 (1962).

114. R. D. Chambers, M. Hole, W. K. R. Musgrave, R. A. Storey, and B. Iddon, *J. Chem. Soc.* **C1966**, 2331 (1966).

115. R. D. Chambers, J. Hutchinson, and W. K. R. Musgrave, *J. Chem. Soc.* **1964**, 3736 (1964).

116. N. Boden, J. W. Emsley, J. Feeney, and L. H. Sutcliffe, *Mol. Phys.* **8**, 133 (1964).

117. R. D. Chambers, J. A. H. McBride, and W. K. R. Musgrave, *Chem. Ind. (London)* **1966**, 904 (1966).

118. D. D. Callander, P. L. Coe, M. F. S. Matough, E. F. Mooney, A. J. Uff, and P. H. Winson, *Chem. Commun.* **1966**, 820.

119. F. Prosser and L. Goodman, *J. Chem. Phys.* **38**, 374 (1963).

120. J. W. Emsley, J. Feeney, and L. H. Sutcliffe, *Prog. in NMR Spectrosc.* **7**, 1 (1971).

Four

^{13}C and Heteroatom Chemical Shifts in Aromatic Compounds

In this chapter we discuss primarily the ^{13}C nuclear magnetic resonance chemical shifts of a variety of polycyclic aromatic hydrocarbons and heterocyclic compounds. The general theory of ^{13}C shieldings has been dealt with in Chapter 2. Thus our discussion here focuses on the basic trends of ^{13}C chemical shifts in these compounds and on the effects of substituents, molecular interactions, and stereochemistry on the chemical shifts. Then, in the last section of this chapter, we round out the discussion of NMR chemical shifts of aromatic compounds by discussing the heteronuclei: ^{15}N, ^{17}O, ^{77}Se, and ^{33}S.

Several books that contain excellent summaries of the ^{13}C NMR literature and have sections dealing with polycyclic aromatic compounds have been published.[1-5] A recent review article by Hansen[6] deals entirely with the ^{13}C NMR of polycyclic aromatic hydrocarbons, gives a thorough survey of the literature on these compounds through mid-1978, and includes extensive tables of chemical shifts. Additionally, ^{13}C spectral data for many aromatic compounds are available on microfiche,[7] and in other compendia of ^{13}C spectra.[8-12]

THE RANGE OF ^{13}C CHEMICAL SHIFTS

Hydrocarbons

The chemical shifts of ^{13}C nuclei cover a broad range, over 200 ppm, or more than 20 times the range of proton chemical shifts. Polycyclic aromatic hydrocarbons have resonances spanning only a fraction of that range: alternant hydrocarbons have a range of less than 10 ppm, while nonalternant hydrocarbons have a somewhat greater range, about 14–22 ppm, reflecting the less uniform distribution of π electrons in the nonalternant compounds. Most of these resonances fall in the general region for aromatic carbons, with δ_C about 118 to 144.[§] Tables 4.1 and 4.2 give δ_C values for some alternant polycyclic hydrocarbons, and Table 4.3 gives δ_C values for some nonalternant compounds.

Substituents on the parent hydrocarbon tend to extend the range of chemical shifts to about 50 ppm. A hydroxyl group, for example, will shift the resonance of the substituted carbon toward high frequency, $\delta_C \sim 160$, reflecting the decreased shielding of the carbon. Near the other extreme of the range, a bromine substituent will shift the resonance of the substituted carbon toward low frequency, $\delta_C \sim 122$.

Heterocycles

The polycyclic aromatic heterocycles have a greater range of ^{13}C chemical shifts than the analogous hydrocarbons. In the series of polycyclic nitrogen heterocycles in Table 4.4, for example, the chemical shifts of the alternant compounds (25–33 and 37) span a range of 40 ppm, from δ_C 121.0 to δ_C 160.7. Inclusion of the nonalternant compounds (34, 35, and 38–42) extends this range further, from δ_C 96.3 for C-1 of 3,4-diazaindene, 41, which is appreciably shielded since it is β to two nitrogens, to δ_C 160.7 for C-2 of quinazoline, 30, which is appreciably deshielded by the adjacent nitrogen. In general, carbons bonded to nitrogen are deshielded, while the β carbons are shielded relative to similar carbons in hydrocarbons.[1]

Analogous trends in the ^{13}C chemical shifts can be seen in the polycyclic compounds and their corresponding monocyclic compounds. Table 4.5 presents a comparison of some of these chemical shifts. The carbons α to the ring junction in naphthalene, 1, and anthracene, 2—C-1, -4, -5, and -8—have approximately the same shielding as benzene. The β carbons and C-9 and -10 in 2 are similar. Carbons at the ring junctions in these hydrocarbons—C-4a, -8a, -9a, and -10a—however, are less shielded by 4–5 ppm, with resonances at δ_C 131.5 and δ_C 133.3.

The nitrogen heterocycles pyridine, 43, quinoline, 25, and acridine, 31, exhibit similar influences, but they are overshadowed in some cases by the influences of the α or β nitrogen. For example, C-2 and -6 in 43, C-2 and -8a in

---

[§] δ_C is the carbon chemical shift in parts per million (ppm) from the tetramethylsilane resonance. A positive δ_C corresponds to a shift to higher frequency (lower field) and thus decreased shielding. Although many literature sources report ^{13}C chemical shift data to 0.01 ppm, because of the various sources (with various solvents, concentrations, etc.) for the data in this chapter, we report these data to only 0.1 ppm.

Table 4.1. ^{13}C NMR Chemical Shifts of Some Alternant Polycyclic Hydrocarbons

Compound	C-1	C-2	C-3	C-4	Other δ_C	Assignment	Reference
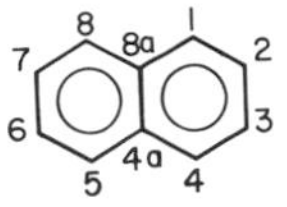 Naphthalene	127.7	125.6	125.6	127.7	133.3	C-4a	13

(1)

Compound	C-1	C-2	C-3	C-4	Other δ_C	Assignment	Reference
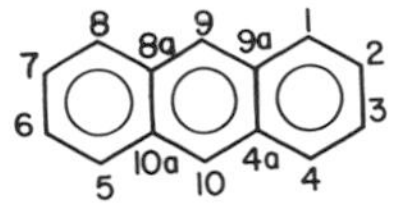 Anthracene	127.9	125.1	125.1	127.9	126.0	C-9	14
					131.5	C-4a	

(2)

Compound	C-1	C-2	C-3	C-4	Other δ_C	Assignment	Reference
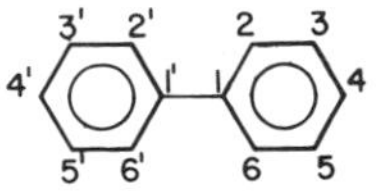 Biphenyl	141.7	127.6	129.2	127.7			15

(3)

Compound	C-1	C-2	C-3	C-4	Other δ_C	Assignment	Reference
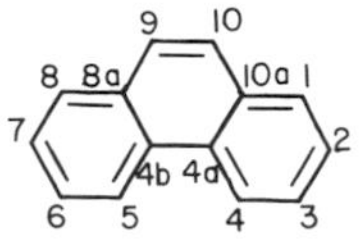 Phenanthrene	128.5	126.5	126.5	122.6	126.9	C-9	16
					130.1	C-4a	
					132.1	C-8a	

(4)

Compound	C-1	C-2	C-3	C-4	Other δ_C	Assignment	Reference
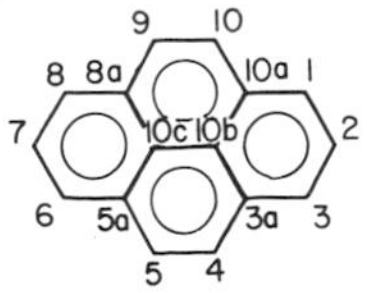 Pyrene	125.4	126.3	125.4	127.9	131.5	C-3a	15
					125.1	C-10b	

(5)

Compound	C-1	C-2	C-3	C-4	Other δ_C	Assignment	Reference
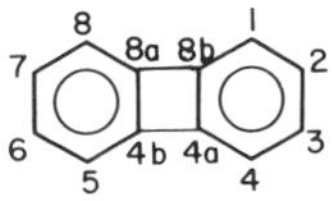 Biphenylene	118.0	128.8	128.8	118.0	151.9	C-4a	17, 18

(6)

Table 4.1. (Continued)

					Other		
Compound	C-1	C-2	C-3	C-4	δ_C	Assignment	Reference
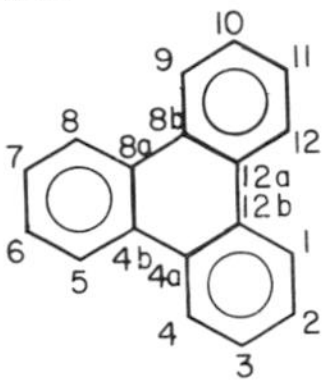 Triphenylene	123.7	127.6	127.6	123.7	130.2	C-4a	17

(7)

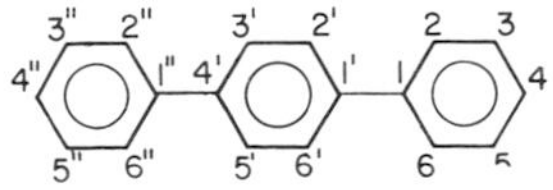 p-Terphenyl	140.8	127.1	128.8	127.4	140.2	C-1′	19
					127.5	C-2′	

(8)

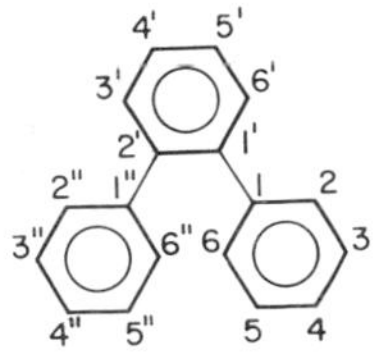 o-Terphenyl	141.6	127.9	129.9	126.4	140.7	C-1′	19
					130.6	C-3′	
					127.5	C-4′	

(9)

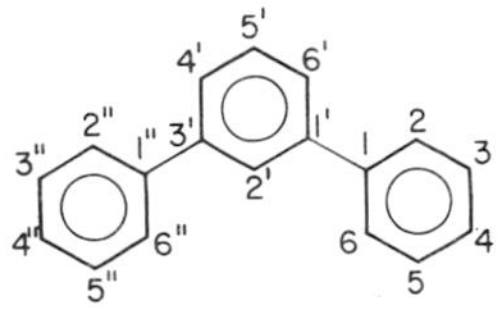 m-Terphenyl	141.3	127.3	128.8	127.4	141.8	C-1′	19
					126.2	C-2′	
					126.2	C-4′	
					129.2	C-5′	

(10)

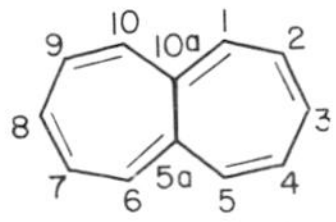 Heptalene (−80°)	137.4	132.8	133.3	132.8	143.1	C-5a	20

(11)

Table 4.1. (Continued)

Compound	C-1	C-2	C-3	C-4	Other δ_C	Assignment	Reference
Pyracyclene	132.4	132.4	124.8	124.8	142.0	C-2a	21
					131.5	C-8b	

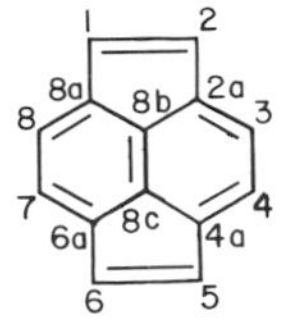

(12)

25, and C-4a and -10a in **31** exhibit a shift downfield by approximately 20 ppm, characteristic of carbons α to a nitrogen. Carbons β to the nitrogen are more shielded than expected, with C-3 in **43** at δ_C 124.5 and C-3 in **25** at 121.7, compared to benzene at δ_C 128.7; β carbons that are also at ring junctions—such as C-4a in **25** at δ_C 128.9 and C-8a and -9a in **43** at δ_C 126.7—are also shielded by the nitrogen, but the shielding is nearly balanced by the characteristic deshielding caused by their presence at the ring junctions.

Little ^{13}C NMR work has been done on polycyclic aromatic compounds with heteroatoms other than nitrogen. Data for several of these are presented in Table 4.6. For comparison, ^{13}C chemical shifts for some monocyclic compounds are also given in the table.

RING-CURRENT EFFECTS

The larger range of chemical shifts observed in the nonalternant, as compared to the alternant, polycyclic aromatic compounds led to the suggestion that the differences in shielding were largely the result of differences in the π-electron density at each carbon.[50,51] Correlations have been obtained between ^{13}C chemical shifts and calculated π-charge densities, with results ranging from about 150 to 210 ppm per π electron.[52] However, deviations from the average of about 160 ppm per π electron in a series of nonalternant hydrocarbons, with the greatest deviations for the quaternary carbons,[25] indicated that other factors must be considered, possibly including ring-current effects. The calculations of Mamaev and his co-workers[53] on ^{13}C shieldings of naphthalene, phenanthrene, and pyrene also indicated a role for ring-current effects in these compounds.[54]

Because the shieldings of ^{13}C nuclei are dominated by the local paramagnetic contributions, as we discussed in Chapter 2, ring-current effects are difficult to

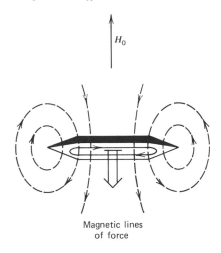

Magnetic lines
of force

Figure 4.1. Current and magnetic lines of force induced in benzene by a primary field, H_0. (From J. A. Pople, W. G. Schneider, and H. J. Bernstein, *High-Resolution Nuclear Magnetic Resonance*, McGraw-Hill, New York (1959).)

observe. The effects of ring currents on the ring carbons can be expected to be small too, as can be seen by examination of the diagram of the magnetic lines of force resulting from induced ring currents in benzene in Figure 4.1.

Alger and his co-workers[15] assumed a diamagnetic ring current in the peripheral ring carbons of pyrene, **5**, to account for the increased shielding of C-10b and -10c in this compound, δ_C 125.1, compared to the more "normal" shieldings of the other quaternary carbons in **5**, δ_C 131.5 for C-3a, -5a, -8a, and -10a. Support for the existence of this ring-current effect arises from the observation of increased shielding of the methyl groups in 10b,10c-dimethyl-10b,10c-dihydropyrene, **53**. Later studies[55,56] compared alkyl chemical shifts in 10b,10c-dialkyl-10b,10c-dihydropyrenes, **54**, in which the alkyl groups should have

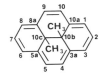

10b,10c-dimethyl-10b,
10c-dihydropyrene
(53)

increased shielding if ring-current effects are important, with shifts in 10b,10c-dialkyl-2,7,10b,10c-tetrahydropyrenes, **55**, in which ring currents should not affect the alkyl chemical shifts. The differences between alkyl chemical shifts in **54** and **55**, both ^1H and ^{13}C, closely followed the theoretical curve of Johnson and Bovey[57] for ring-current effects.

Table 4.2. ^{13}C NMR Chemical Shifts of Some Alternant Polycyclic Hydrocarbons

Compound	C-1	C-2	C-3	C-4	C-5	C-6	C-7	C-8
Benzo[a]pyrene	124.6	125.8	124.7	127.6	128.0	125.4	128.8	125.88

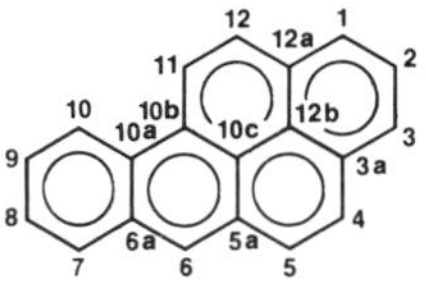

(13)

Benz[a]anthracene	122.8	126.7	126.6	128.5	126.9	126.9	127.2	128.3

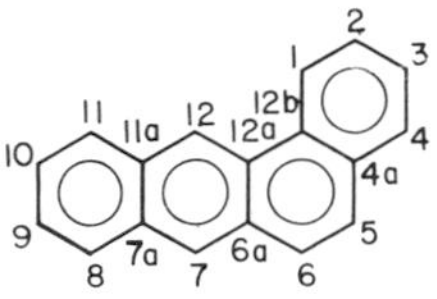

(14)

Dibenz[a,h]anthracene	122.8	126.9	126.7	128.6	127.1	127.4	122.1	122.8

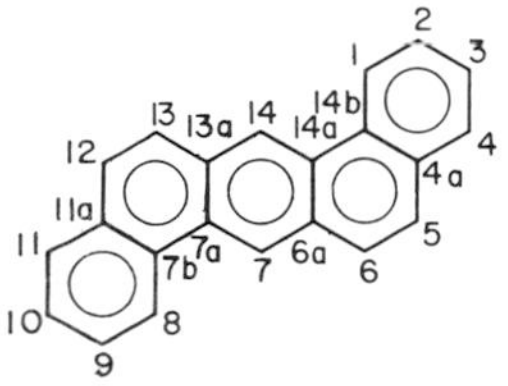

(15)

Dibenz[a,c]anthracene	123.4	127.4	127.6	123.6	123.6	127.6	127.4	123.4

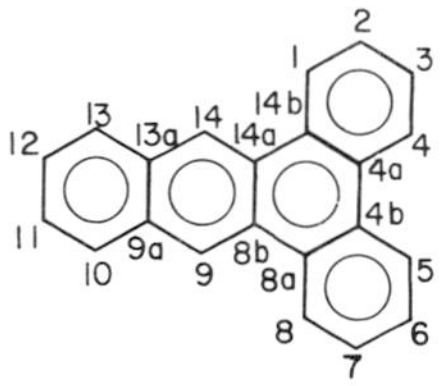

(16)

Table 4.2. (Continued)

C-9	C-10	C-11	C-12	C-13	C-14	Other δ_C	Assignment	Reference
125.91	122.9	122.0	127.3			131.2	C-3a	22, 23
						129.7	C-5a	
						131.5	C-6a	
						127.2	C-10a	
						128.1	C-10b	
						131.3	C-12a	
						125.3	C-12b	
						123.6	C-12c	
125.6	125.7	127.6	121.4			131.9	C-4a	24
						131.9	C-6a	
						131.8	C-7a	
						128.8	C-11a	
						130.5	C-12a	
						130.6	C-12b	
126.9	126.7	128.6	127.1	127.4	122.1	132.0	C-4a	24
						129.1	C-6a	
						130.8	C-7a	
						130.2	C-7b	
122.0	128.1	126.0	126.0	128.1	122.0	130.1	C-4a	24
						130.2	C-8a	
						128.5	C-8b	
						132.3	C-9a	

Table 4.3. ^{13}C NMR Chemical Shifts of Some Nonalternant Polycyclic Hydrocarbons

Compound	C-1	C-2	C-3	C-4	C-5	C-6	C-7	C-8	Other		
									δ	Assignment	Reference
Azulene 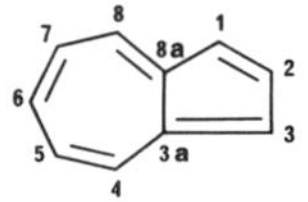 (17)	119.2	137.9	119.2	136.9	123.2	137.4	123.2	136.9	140.8	C-3a	25, 26, 27
Acenaphthylene 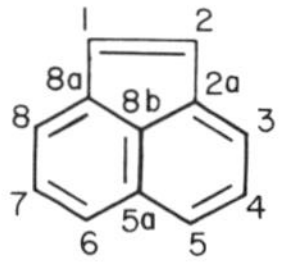 (18)	129.9	129.9	124.5	128.0	127.5	127.5	128.0	124.5	140.2 128.9 128.6	C-2a C-5a C-8b	27
Fluoranthene 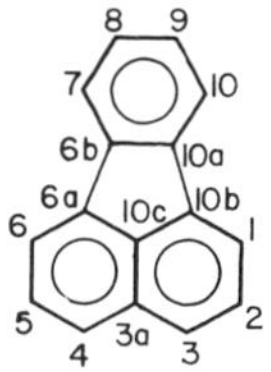 (19)	121.0	128.8	127.4	127.4	128.8	121.0	122.3	128.4	130.9 137.6 140.1 132.9	C-3a C-6a C-6b C-10c	27, 28
Fluorene 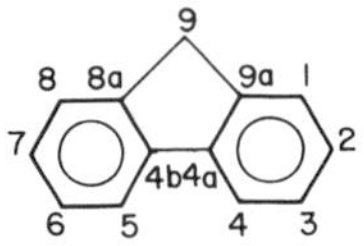 (20)	125.0	126.7	126.7	119.8	119.8	126.7	126.7	125.0	141.7 143.2 36.9	C-4a C-8a C-9	16

Azulene (17)

Acenaphthylene (18)

Fluoranthene (19)

Fluorene (20)

Compound									Quaternary carbons	Ref.
Acenaphthene 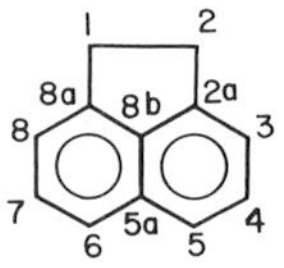 (21)	30.4	30.4	119.6	128.4	122.9	122.9	128.4	119.6	146.1 C-2a 132.3 C-5a 139.9 C-8b	27
Acepleiadiene 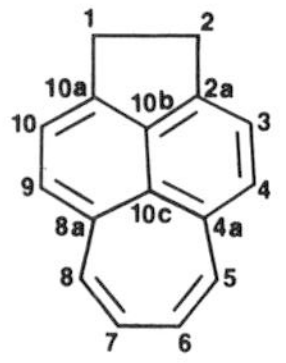 (22)	29.6	29.6	120.0	127.9	125.9	138.4	138.4	125.9	143.9 C-2a 135.8 C-4a 142.9 C-10b 136.5 C-10c	29
Acepleiadylene 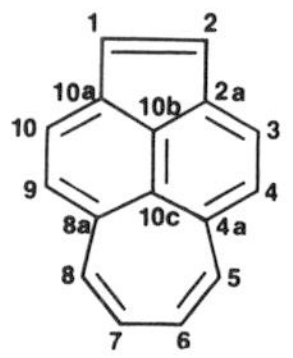 (23)	126.1	126.1	125.8	127.4	126.8	137.0	137.0	126.8	138.2 C-2a 134.8 C-4a 126.6 C-10b 127.2 C-10c	29
Pleiadiene 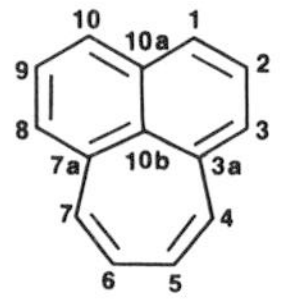 (24)	128.83	128.76	127.7	127.1	139.9	139.9	127.1	127.7	139.9 C-3a 138.9 C-10a 139.2 C-10b	30

Acenaphthene (21)

Acepleiadiene (22)

Acepleiadylene (23)

Pleiadiene (24)

Table 4.4. ^{13}C NMR Chemical Shifts of Some Polycyclic Aromatic Heterocycles

Compound	C-1	C-2	C-3	C-4	C-5	C-6
Quinoline 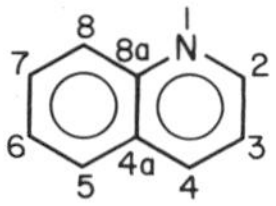 (25)	—	151.1	121.7	136.2	128.5	127.0
Isoquinoline 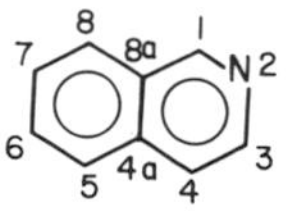 (26)	153.3	—	144.0	121.0	127.0	130.7
Cinnoline 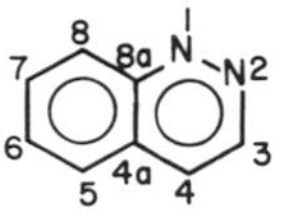 (27)	—	—	146.3	124.8	128.1	132.5
Phthalazine 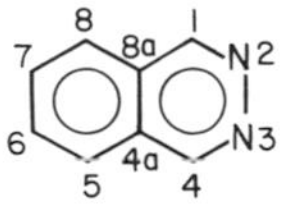 (28)	152.1	—	—	152.1	126.9	133.4
Quinoxaline 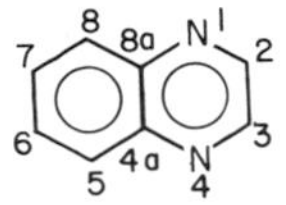 (29)	—	145.7	145.7	—	130.0	130.1
Quinazoline 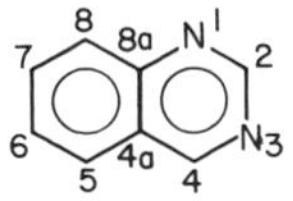 (30)	—	160.7	—	155.9	127.6	128.1
Acridine 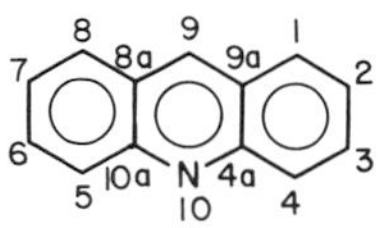 (31)	130.5	125.7	128.5	129.7	129.7	128.5
Phenazine 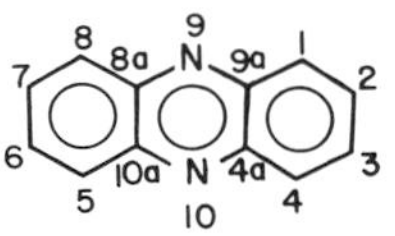 (32)	130.2	129.6	129.6	130.2	130.2	129.6

Table 4.4. (Continued)

C-7	C-8	C-9	C-10	C-4a	C-8a	δ	Assignment	Reference
						Other		
129.9	130.3			128.9	149.1			1, 31, 32, 33
127.7	128.1			136.2	129.3			1, 31
132.3	129.7			127.0	151.2			1, 31
133.4	126.9			126.9	126.9			1, 31
130.1	130.0			143.4	143.4			1, 31, 33
134.3	128.8			125.4	150.3			1, 31
125.7	130.5	136.1	—	149.3	126.7			1, 31
129.6	130.2	—	—	143.5	143.5			34

Table 4.4. (Continued)

Compound	C-1	C-2	C-3	C-4	C-5	C-6
Pteridine 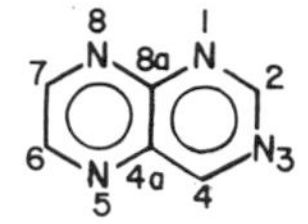	—	163.7	—	159.0	—	147.8
Purine 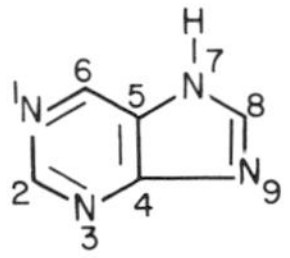	—	151.8	—	154.6	128.3	144.6
Indazole 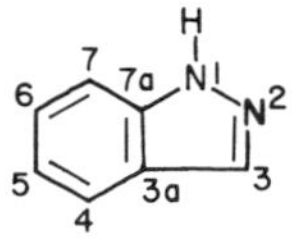	—	—	133.4	120.4	120.1	125.8
Oxazolidine 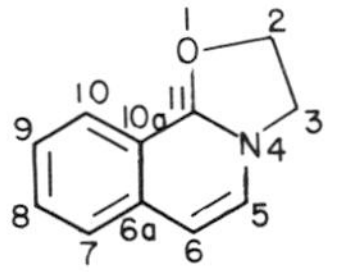	—	62.3	59.3	—	130.5	102.5
1,10-Phenanthroline 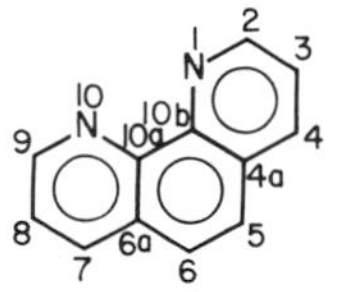	—	150.6	123.4	136.3	126.8	126.8
4-Azaindene 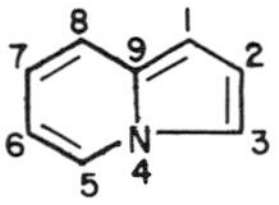	99.4	114.1	113.0	—	125.6	110.4
1,4-Diazaindene 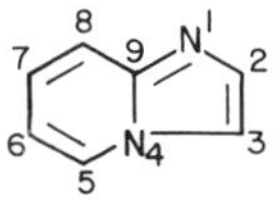	—	134.1	113.4	—	127.0	112.2

Pteridine (33)

Purine (34)

Indazole (35)

Oxazolidine (36)

1,10-Phenanthroline (37)

4-Azaindene (38)

1,4-Diazaindene (39)

Table 4.4. (Continued)

C-7	C-8	C-9	C-10	C-4a	C-8a	Other δ	Assignment	Reference
152.4	—			135.0	152.4			35, 36
—	147.7	—						37, 38, 39
110.0						122.8	C-3a	40, 41
						139.9	C-7a	
123.7	128.9	124.8	133.9			134.4	C-6a	42
						121.3	C-10a	
						76.1	C-11	
136.3	123.4	150.6	—	129.0		145.6	C-10a	43
117.2	119.6	133.4						44
124.6	117.6	145.6						44

Table 4.4. (Continued)

Compound	C-1	C-2	C-3	C-4	C-5	C-6
2,4-Diazaindene 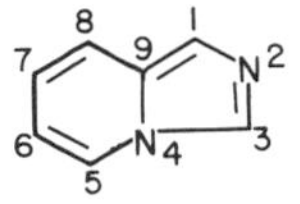	119.9	–	128.4	–	122.8	112.7
3,4-Diazaindene 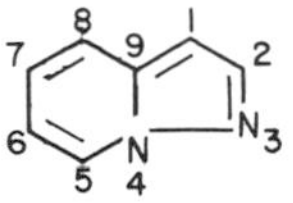	96.3	141.3	–	–	128.1	110.8
1,4,8-Triazaindene 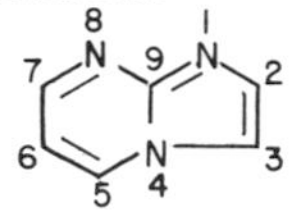	–	135.1	112.7	–	136.0	109.4

(40)

(41)

(42)

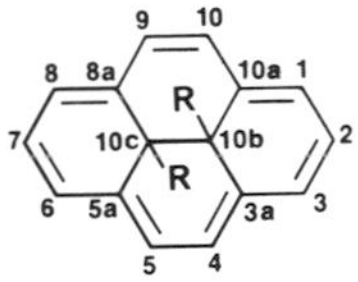

10b,10c-dialkyl-10b,
10c-dihydropyrenes
(54)

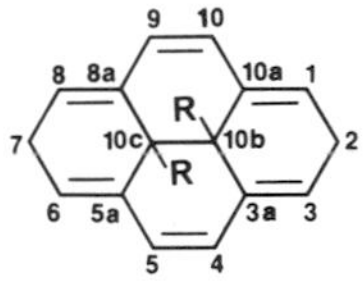

10b,10c-dialkyl-2,7,10b,
10c-tetrahydropyrenes
(55)

Since the effects of ring currents on ^{13}C shieldings are small, they are easily covered up by other effects, such as those of ring strain or conformational changes. Thus, Günther and his co-workers[58] concluded from their study of bridged annulenes that diamagnetic ring currents can be detected, but that ^{13}C NMR is not a reliable method for measurement of ring-current effects in aromatic systems.

In certain cases, though, ^{13}C NMR can be highly effective as a probe of ring-current effects. For a series of pyracyclene, **12**, derivatives, evidence was found for a paramagnetic ring-current shift of about 8 ppm for the central carbons C-8b and -8c. Similarly, acepleiadylene, **23**, was estimated to have a diamagnetic ring-current contribution to the shieldings of the central carbons C-10b and -10c of about 5.4 to 7.3 ppm.[21] It appears that ^{13}C NMR is useful for examination of ring-current effects in some series of compounds, but its general validity for this examination remains to be established.

Table 4.4. **(Continued)**

| | | | | | | Other | | |
C-7	C-8	C-9	C-10	C-4a	C-8a	δ	Assignment	Reference
119.4	118.2	130.6						44
122.4	117.4	139.5						44
150.9	–	148.9						44

STERIC AND CONFORMATIONAL EFFECTS

It is well established that molecular stereochemistry can have significant effects on the ^{13}C shieldings of alkanes, alkenes, and various alicyclic molecules.[1,59] It is also well established that the effects of substituents on the ^{13}C shieldings of aromatic molecules can be altered markedly by stereochemical interactions, as will be discussed later in this chapter. Examples in the literature of steric effects on the ^{13}C NMR spectra of the unsubstituted polycyclic aromatic compounds, however, are scarce, as might be expected for these (mostly) planar molecules. One apparent example of a steric effect on the ^{13}C shieldings of the unsubstituted polycyclic aromatic hydrocarbons is the ^{13}C NMR spectrum of phenanthrene, **4**. The carbons in the bay positions, C-4 and -5, have a ^{13}C NMR chemical shift of δ_C 122.6, indicating significant shielding of these carbons compared to the relatively noncrowded carbons in benzene, δ_C 128.7; C-1, -4, -5, and -8 in naphthalene, **1**, δ_C 127.7; C-1, -4, -5, and -8 in anthracene, **2**, δ_C 127.9; or C-2, -6, -2', and -6' in biphenyl, **3**, δ_C 127.6. In biphenyl, essentially free rotation about the inter-ring bond occurs at ambient temperatures.[60] The slight peri interaction in **1** and **2** does not appear to have a significant effect on the ^{13}C chemical shifts.

Grant and Cheney[61] proposed that an upfield shift of the ^{13}C resonance should result from steric perturbation of C–H bonds, where the neighboring hydrogens are sufficiently close that charge polarization is induced in the interacting bonds by nonbonded repulsions between the hydrogens. In phenanthrene,

Table 4.5. Comparison of ^{13}C Chemical Shifts of Some Monocyclic and Polycyclic Hydrocarbons and Heterocycles

Compound	C-1	C-2	C-3	C-4	C-5	C-6	C-7	C-8	C-9	C-10	C-4a	C-8a	C-9a	C-10a
Benzene	128.7	128.7	128.7	128.7	128.7	128.7								
Naphthalene, 1	127.7	125.6	125.6	127.7	127.7	125.6	125.6	127.7			133.3	133.3		
Anthracene, 2	127.9	125.1	125.1	127.9	127.9	125.1	125.1	127.9	126.0	126.0	131.5	131.5	131.5	131.5
Pyridine	—	150.6	124.5	136.4	124.5	150.6								
Quinoline, 25	—	151.1	121.7	136.2	128.5	127.0	129.9	130.3			128.9	149.1		
Acridine, 31	130.5	125.7	128.5	129.7	129.7	128.5	125.7	130.5	136.1	—	149.3	126.7	126.7	149.3

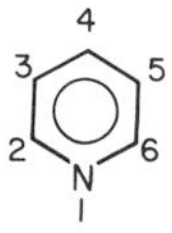

(43)

Table 4.6. ^{13}C NMR Chemical Shifts of Some Heterocyclic Aromatic Compounds

Compound	C-1	C-2	C-3	C-4	C-9	C-4a	C-9a	C-10a	Other	Reference
Pyrrole 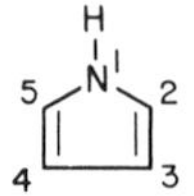 **(44)**	—	118.7	108.4	108.4						1, 45
Furan 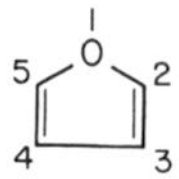 **(45)**	—	142.8	109.8	109.8						
Thiophene 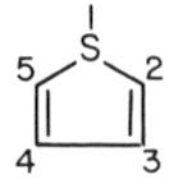 **(46)**	—	125.6	127.4	127.4						
Selenophene 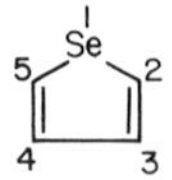 **(47)**	—	130.5	129.2	129.2						46
Phenothiazine 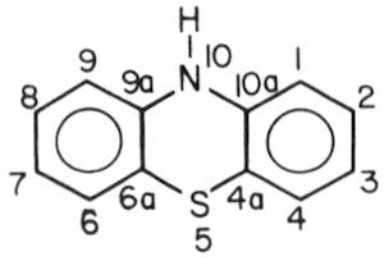 **(48)**	114.3	127.3	121.6	126.0	114.3	116.2	142.0	142.0		47

Table 4.6. (Continued)

Compound	C-1	C-2	C-3	C-4	C-9	C-4a	C-9a	C-10a	Other	Reference
2,3-Diazaphenothiazine 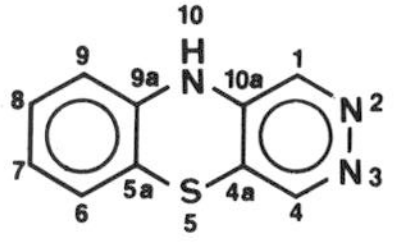 (49)	137.4	—	—	146.2	115.6	117.0	138.4	139.5	C-6 126.6 C-7 123.5 C-8 128.0 C-5a 114.1	47
Xanthene 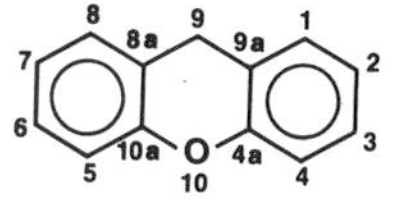 (50)	128.9	123.0	127.7	116.6	27.9	152.1	120.8			48
Benzothiophene 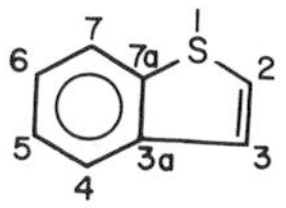 (51)	—	122.6	126.3	124.3					C-3a 139.9 C-7a 140.1 C-5 123.8 C-6 123.9 C-7 124.4	49
Dibenzothiophene 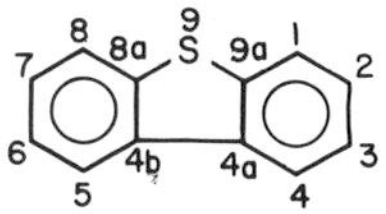 (52)	126.8	122.9	121.8	124.4	—	135.9	139.8			49

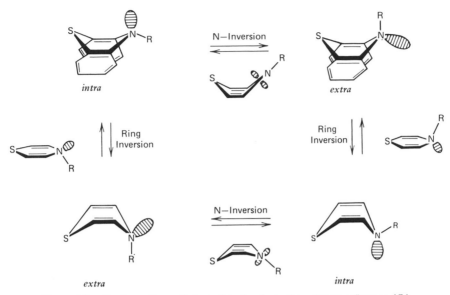

Figure 4.2. Conformations of phenothiazine derivatives. (From reference 47.)

these nonbonded interactions can certainly occur between H-4 and H-5, which lie only ~1.75 Å apart, far less than the sum of their Van der Waals radii.[1]

Fronza and his colleagues[47] have studied the electronic properties and conformations of phenothiazine, **48**, 2,3-diazaphenothiazine, **49**, and several of their derivatives. Both these compounds have a nonplanar structure, with the phenothiazine central ring folded along the S–N axis. Inversion of the central ring or inversion about the nitrogen can take place, and other techniques (electron spin resonance, polarographic oxidation potentials) suggested that the parent compounds have the *intra* conformation, whereas the 10-substituted derivatives, **56** and **57**, have the *extra* conformation, as shown in Figure 4.2. The NMR data

10-substituted phenothiazines
(56)

10-substituted
2,3-diazaphenothiazines
(57)

showed that both inversion processes are rapid above -60°C and confirmed the *intra* conformation as the preferred one. With a heavy ligand R, with peri crowding, the population of the *extra* conformer appears to increase, as is shown by

Table 4.7. ^{13}C NMR Chemical Shifts of Some Phenothiazines

Compound	C-1	C-2	C-3	C-4	C-6	C-7	C-8	C-9	C-4a	C-5a	C-9a	C-10a	CH$_3$	Reference
Phenothiazine, 48	114.3	127.3	121.6	126.0	126.0	121.6	127.3	114.3	116.2	116.2	142.0	142.0	–	47
10-Methylphenothiazine, 56	114.6	127.8	122.5	126.8	126.8	122.5	127.8	114.6	122.4	122.4	145.4	145.4	35.2	47
2,3-Diazaphenothiazine, 49	137.4	–	–	146.2	126.6	123.5	128.0	115.6	117.0	114.1	138.4	139.5	–	47
10-Methyl-2,3-diazaphenothiazine, 57	137.8	–	–	146.7	127.5	124.2	128.5	115.6	122.0	118.8	(142.7)	(142.5)	33.9	47

the deshielding effect of methylation at position 10 on carbons in the benzene and pyridazine rings in **56** and **57** (R = CH_3) relative to **48** and **49**. The *extra* conformer is expected to be deshielded relative to the *intra*, particularly at the ortho and para positions, because of the decreased delocalization of the nitrogen lone pair electrons in the *extra* conformer. Chemical shift data for **48** and **49** (R = H) and **56** and **57** (R = CH_3) are given in Table 4.7.

Recently, Goutarel and his co-workers[62] have derived conformational information for several indole alkaloids from ^{13}C NMR data.

SUBSTITUENT EFFECTS

Charge Density Dependences

Because of the dominance of the local paramagnetism in ^{13}C shieldings, ^{13}C chemical shifts are quite sensitive to changes in the electronic charges at the carbon.[1,52] Introduction of a substituent in the molecule affects the distribution of charge at the various ^{13}C nuclei, which, in turn, affects the ^{13}C chemical shifts. Additionally, since carbon nuclei are on the "inside" of the molecule, rather than on the periphery, and subject to a variety of subtle *inter*molecular influences, as are protons, ^{13}C chemical shifts reflect more directly than do ^1H chemical shifts the effects of substituents on the basic properties of the molecule. Substituent-induced chemical shifts, or "substituent effects," thus play a more important role in ^{13}C than in ^1H NMR spectroscopy. The importance of this role is increased by two other advantages of the ^{13}C nucleus as a probe of molecular properties—a large range of chemical shifts, greater than 200 ppm, and a small or negligible perturbation of the molecular environment relative to its more abundant nonmagnetic isotope ^{12}C.

The simplest systems for the study of substituent effects in cyclic aromatic compounds are monosubstituted benzenes. Data for over 700 of these compounds have been collected in a comprehensive critical review by Ewing.[63]

Pi and Sigma Electronic Effects

Correlations between ^{13}C chemical shifts and electronic charge distributions for monosubstituted benzenes were sought by research groups led by Nelson[64] and later by Olah.[65] These correlations were good for the total charge density and the π-charge density for carbons para to the substituent, as shown in Figure 4.3, but not so good for ortho or meta carbons.

Several papers describe molecular orbital calculations of electronic charge distributions in methyl-substituted polycyclic aromatic hydrocarbons and relate

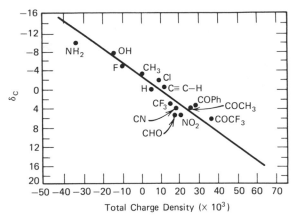

Figure 4.3. ^{13}C chemical shifts of monosubstituted benzenes versus total charge densities calculated by the CNDO/2 method. (From reference 64.)

these charge distributions to the ^{13}C chemical shifts. For 1- and 2-methylnaphthalene, **58** and **59**, and 1,2-, 1,3-, and 1,8-dimethylnaphthalene, **60, 61,** and **62,**

1-methylnaphthalene
(58)

2-methylnaphthalene
(59)

1,2-dimethylnaphthalene
(60)

1,3-dimethylnaphthalene
(61)

1,8-dimethylnaphthalene
(62)

respectively, Wilson and Stothers[13] found that CNDO/2[66] π-charge densities had a marginally better correlation with δ_C for the quaternary carbons than did the total charge densities; overall, the total charge gave better correlations, but neither was exceptionally good, as can be seen by examination of Figure 4.4.

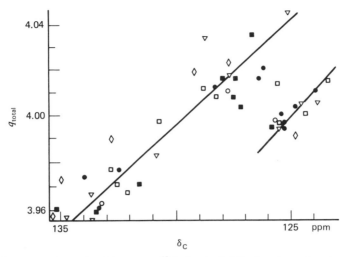

Figure 4.4. Total charge densities versus ^{13}C chemical shifts for the aryl carbons in naphthalene (○) and its 1-methyl (●), 2-methyl (■), 1,2-dimethyl (□), 1,3-dimethyl (▽), and 1,8-dimethyl (◇) derivatives. (From reference 13.)

Aryl carbons in the β positions of unsubstituted rings and C-8 in 1-substituted rings are more shielded than the trend for the other carbons would indicate. These β carbons correspond to the points about the lower line in Figure 4.4.

At first examination, one might suspect that inclusion of methylnapthalenes with possibly significant methyl–methyl steric interactions—the ortho interaction in 1,2-dimethylnaphthalene and the peri interaction in 1,8-dimethylnaphthalene—was largely responsible for the scatter in Figure 4.4. However, a later study[14] of methylated anthracenes in which only compounds lacking ortho or peri disubstitution (1-, 2-, and 9-methyl-; 1,4-, 1,8-, and 9,10-dimethyl-; 2,7,9-trimethyl-; and 1,4,5,8-tetramethylanthracene; **63–70**) were included in the

1-methylanthracene
(63)

2-methylanthracene
(64)

9-methylanthracene
(65)

1,4-dimethylanthracene
(66)

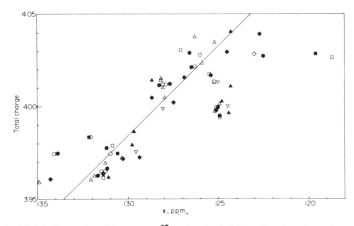

1,8-dimethylanthracene
(67)

9,10-dimethylanthracene
(68)

2,7,9-trimethylanthracene
(69)

1,4,5,8-tetramethylanthracene
(70)

correlation gave a similar correlation between δ_C and CNDO/2 charge densities, as is shown in Figure 4.5. Thus, although the ^{13}C shieldings are related to the charge densities, other contributions, including steric and anisotropy effects, must be taken into account.

Substituent effects in 1- and 2-monosubstituted naphthalenes were investigated by Ernst[67] and correlated with electronic charge densities calculated by the INDO[66] approximation. The substituents—R = NO_2, CN, $COCH_3$, COOH, CHO, F, OCH_3, OH, CH_3, and NH_2—covered a relatively wide range of electron-donating and electron-releasing potentials. Thus, these correlations are with a much wider range of charge densities and chemical shifts and could be expected

Figure 4.5. Total charge densities versus ^{13}C chemical shifts of aryl carbons in anthracene (o) and its 1-methyl (●), 2-methyl (△), 9-methyl (▲), 1,4-dimethyl (◇), 2,7,9-trimethyl (♦), 1,8-dimethyl (□), 1,4,5,8-tetramethyl (■), and 9,10-dimethyl (▽) derivatives. (From reference 14.)

to give better indications of trends than those for the isomeric methylnaphthalenes discussed above, and they are believed to be less affected by minor steric and other perturbations. Good linear dependences were found in some cases—notably for C-4 and C-7 in α-naphthalenes, **71**, and for C-6, -8, and -4a in β-naphthalenes, **72**.

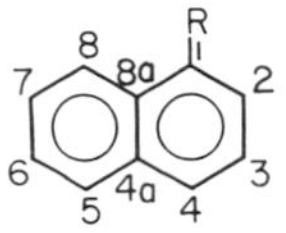

1-substituted
naphthalenes
(71)

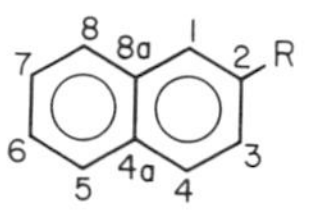

2-substituted
naphthalenes
(72)

Hückel and extended Hückel molecular orbital theory have been used[68] to calculate electron distributions in methyl derivatives of azulene, **17**; they have been related to the δ_C values with modest success. Marker and his co-workers[43] have recently studied charge distributions calculated by the CNDO/2 technique and the ^{13}C chemical shifts in 1,10-phenanthroline, **37**, and some of its derivatives.

Charge densities from CNDO/2 calculations have also been correlated with δ_C values for a series of 4,4′-disubstituted biphenyls,[69] **73**—R = NO_2, Cl, F,

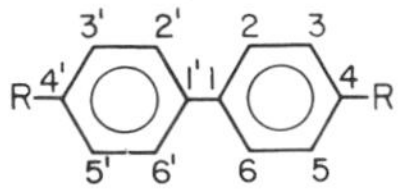

4,4′-disubstituted biphenyls
(73)

CH_3O, CH_3, and H—with good results for total charge densities versus δ_C. In the same work by Wilson and Anderson,[69] the CNDO/2 charge densities are compared with the ^{13}C shieldings for a number of symmetrical chlorinated biphenyls. It is shown that the σ-charge distribution, rather than the π-charge distribution, is the main determinant of differences in δ_C at various positions for different patterns and degrees of chlorine substitution, as is shown in Figure 4.6.

Thus, although changes in the π-charge distribution brought about by substitution on an aromatic ring make a major contribution to the changes in chemical shifts, changes in σ-charge distribution also influence the ^{13}C shieldings and can assume major importance for substituents (like chlorine) that have a small mesomeric effect. Later, Wilson[60] attempted to relate CNDO/2 charge distributions as a function of inter-ring dihedral angle for sterically hindered chlorinated biphenyls (those with ortho–ortho′ substitution—for example, 2,2′-dichlorobiphenyl, **74**, and 2,2′,6,6′-tetrachlorobiphenyl, **75**) to the ^{13}C chemical shifts at

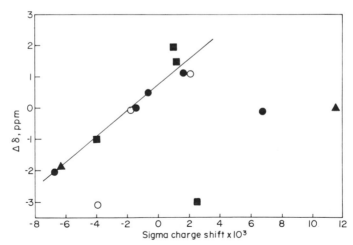

2,2'-dichlorobiphenyl
(74)

various temperatures. The attempt was unsuccessful but was felt to be a failure of the CNDO/2 technique for highly hindered geometries in these systems. The calculated minimum energy dihedral angles, about 26°, and potential barriers, about 150-200 kcal/mol, were somewhat unrealistic. In **75**, for example,

2,6,2',6'-tetrachlorobiphenyl
(75)

approach of the ortho-ortho' chlorines within the sum of their Van der Waals radii would allow only a much larger dihedral angle, about 58°.

In the work by the Fronza group[47] on phenothiazine derivatives described earlier in this chapter, CNDO/2 calculations of π and total charge densities were

Figure 4.6. Shift of the ^{13}C shielding from biphenyl versus the calculated σ charge shift from biphenyl for freely rotating chlorinated biphenyls. ● 3,3'; ▲ 3,5,3',5'; ■ 3,4,3',4'; ○ 4,4'. Points far below the line correspond to C-1 and C-1'. (From reference 69.)

performed for phenothiazine, **48**, 2,3-diazaphenothiazine, **49**, and 1-chloro-2, 3-diazaphenothiazine, **76**, using geometries from the X-ray structures. A comparison of the calculated π-electron densities to ^{13}C chemical shifts is shown in

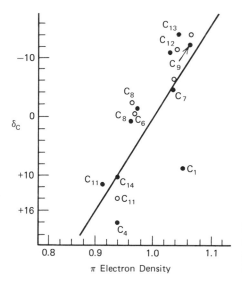

1-chloro-
2,3-diazaphenothiazine
(76)

Figure 4.7. The straight line shown is for a slope of 160 ppm per unit electronic charge, the average value suggested by Spiesecke and Schneider.[25] The scatter is large and is greater for carbons adjacent to heteroatoms—in particular, C-1 and C-4. A similar correlation was found for total charge density and ^{13}C chemical shifts. In this case, it appears that the charge density contribution to the shieldings, although important, is not dominant, and the chemical shifts must be interpreted with inclusion of other effects, such as the anistropy of the adjacent nitrogen atoms, or electric field effects.

Witanowski and his colleagues[70] have examined the ^{13}C chemical shifts of a number of azaindolizines, which contain all possible combinations of nitrogen atoms within the five-membered ring moiety of indolizine, **77**. Linear correla-

Figure 4.7. Calculated π-charge densities versus ^{13}C chemical shifts for phenothiazine (●) and 2,3-dibenzophenothiazine (○). (From reference 47.)

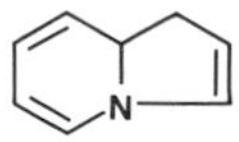

indolizine
(77)

tions were sought between the δ_C values and total, rather than π, charge densities calculated by the INDO method. Although the π-charge distributions were as expected from consideration of simple resonance structures, the chemical shift correlations depended significantly on the sigma electronic polarization. This result is similar to those obtained earlier for chlorinated biphenyls by Wilson and Anderson.[69]

Correlations with Mesomeric and Inductive Substituent Parameters

As we discussed above, the major effects of a substituent on ^{13}C chemical shifts must be related to the ability of the substituent to change the electronic environment at the ^{13}C nuclei of interest. These effects are transferred to a site in the molecule by inductive processes, mesomerism or resonance, and direct field effects. The first set of parameters that described substituent effects on reactivity was developed by Hammett[71] and has been refined over the years.[72] The Hammett equation, in a form appropriate to NMR, assumes a simple proportionality between the effect, $\Delta\delta$, of a substituent on the ^{13}C chemical shift and the appropriate Hammett parameter, σ.

$$\Delta\delta = \rho\sigma \qquad (4.1)$$

The choice of σ depends on the type of substituent. Hammett parameters have been correlated successfully with ^{13}C chemical shifts in mono-substituted benzenes,[64] and used somewhat less successfully[73] to analyze ^{19}F chemical shifts in polynuclear aromatic compounds.

Several variations on the Hammett substituent parameters have been developed. These include the variable combination of field, $\mathcal{F}$, and resonance, $\mathcal{R}$, parameters unique to the substituent, which are weighted by empirical coefficients f and r to give the substituent parameters

$$\sigma = f\mathcal{F} + r\mathcal{R} \qquad (4.2)$$

suggested by Swain and Lupton.[74] The relative magnitudes of r and f give the percentages of resonance character of a given parameter; for instance, σ_p^+ reflects approximately 66% resonance, and σ_p reflects 53% resonance, but σ_m reflects approximately 22% resonance contributions.

The Taft dual substituent parameter (DSP) approach has also been used to separate the inductive and mesomeric influences of substituents. The dual substituent parameter equation separates substituent effects into inductive ($\rho_I \sigma_I$) and resonance ($\rho_R \sigma_R$) components.[75]

$$\Delta\delta = \rho_I \sigma_I + \rho_R \sigma_R \qquad (4.3)$$

One of four different resonance scales—$\sigma_R = \sigma_R^-$, σ_R^0, σ_R^{BA}, or σ_R^+—is used, depending on the electron demand at the measuring site.[76] Bromilow and Brownlee[77] have discussed the advantages of using the DSP method in preference to a single parameter scale for analysis of ^{13}C substituent chemical shift data. Using a graphical representation of the DSP equation (4.3), they developed a recommended basis set of substituents. This recommended set is the following: two strong donors [$N(CH_3)_2$, NH_2, or OCH_3]; two halogens (not both Cl and Br); CH_3 and H; one acceptor (NO_2, CN, or CF_3); and one carbonyl acceptor ($COCH_3$ or COOR).

Dewar[78] has used a three-parameter approach (*FMMF*) with a weighted combination of field, F, mesomeric, M, and mesomeric field, MF, parameters. The coefficients reflect the molecular geometry. The limitations of this method have been discussed by Reynolds and Hamer.[79]

Several applications of these substituent parameter analyses to the interpretation of ^{13}C NMR data on polycyclic aromatic molecules are in the literature. Schulman and his co-workers[80] studied ^{13}C chemical shifts in 14 mono-substituted benzenes and 14 4-substituted biphenyls, **78**, with the aim of finding the

4-substituted biphenyls
(**78**)

best σ parameters of those in the literature for correlation and prediction of the chemical shifts. Using linear regression analysis and various models, they found good correlations between the substituent effects on the δ_C values, $\Delta\delta_C$, and the appropriate σ parameters for the 3 and 4 positions of benzene and the 1, 2, 1′, 2′, 3′, and 4′ positions of biphenyl. The 1 and 2 positions of benzene and the 4 position of biphenyl did not give good correlations, presumably because of the local and steric effects of the substituents. Positions 4 in benzene and 1 in biphenyls gave the best fits with σ_p^+, and positions 3 in benzene and 2, 1′, and 3′ in biphenyls gave the best fits with σ_m. For C-3 of benzene and C-2 of biphenyl a plot of the substituent effect $\Delta\delta_C$ vs. σ_m^+ gave a good fit, but even better results were obtained when the data for substituents with and without lone pair electrons were plotted separately, as in Figure 4.8. A surprising feature of this work was the finding that the effect of a 4-substituent in biphenyl was transmitted (with a 0.99 correlation) to the 4′ carbon through eight covalent bonds.

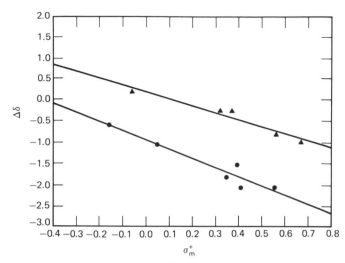

Figure 4.8. Substituent effect, $\Delta\delta$, on the ^{13}C chemical shift of the 2-carbon of 4-substituted biphenyls versus σ_m^+. ▲, Substituents with unshared electron pairs; •, substituents without unshared electron pairs. (From reference 80.)

Recently, Wilson[81,82] has examined the ^{13}C chemical shifts of 4-substituted paraterphenyls, **79,** and, using the Hammett and Taft equations, found good

4-substituted paraterphenyls
(79)

correlations with the sigma parameters at most positions. A significant substituent effect was observed even at C-4″, which is 12 covalent bonds away from the substituent! One wonders if C-4‴ in a 4-substituted paraquaterphenyl, **80,**

4-substituted paraquaterphenyls
(80)

would also sense the presence of a 4-substituent. This effect seems reminiscent of the tale of the princess and the pea. Work on terphenyls is discussed in more detail later in this chapter.

Using the Dewar *FMMF* treatment, Schulman and his colleagues[80] obtained independent correlations of $\Delta\delta_C$ with σ for each position in the monosubstituted benzenes and 4-substituted biphenyls quite similar to the more classical correlations described above. Their analysis implied that the mesomeric, *M*, effects are 41 times greater, and the mesomeric field, *MF*, effects four times greater than the field, *F*, effects in these compounds. The limitations[79] of the *FMMF* method must be kept in mind in interpreting these results. In any event, the importance of resonance contributions to ^{13}C shieldings is thus underlined.

Of the several approaches outlined above, the most useful seems to be the dual substituent parameter approach, although occasional successes of a simple Hammett approach are reported,[81] as in a study of 19 6- and 7-substituted coumarins,[83] or the study of substituted purines by the Thorpe group.[39] Shapiro[84] used the Swain–Lupton parameters to analyze substituent shifts at the meta and para carbons in a series of 9-substituted fluorenes, **81**, and 9-substituted 1-methylfluorenes, **82**. The analysis suggested that π-inductive effects were twice as

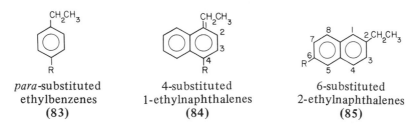

9-substituted fluorenes (81) 9-substituted 1-methylfluorenes (82)

important as hyperconjugative interactions. However, because of the small number and similarity of the substituents studied (R = OH, Cl, Br, and I), the results should be viewed with caution, particularly in a quantitative sense.

The Taft dual substituent parameters (DSP) were used by groups led by Wells[75] and Ehrenson[76] to analyze ^{13}C chemical shift data for several aromatic systems. Adcock and his colleagues[85] analyzed the effects of para substituents in ethylbenzenes, **83**, 4-substituents in 1-ethylnaphthalenes, **84**, and 6-substituents in 2-ethylnaphthalenes, **85**, on the ^{13}C shieldings of the ethyl and ipso

para-substituted ethylbenzenes (83)   4-substituted 1-ethylnaphthalenes (84)   6-substituted 2-ethylnaphthalenes (85)

carbons with the DSP treatment. "Normal" behavior—that is, upfield shifts resulting from electron donor groups and downfield shifts resulting from electron acceptor groups—was observed for the methylene and the ipso carbons, whereas "inverse" behavior was observed for the methyl carbons. Substituent shifts appeared to be more sensitive to resonance phenomena in the naphthyl compounds, but no clear-cut generalization could be made about the relative importance of inductive and resonance phenomena in the three systems.

A study by Kitching and his colleagues[86] appeared at about the same time. It used a fairly large set of 1- and 2-substituted naphthalene (**71** and **72**) spectra obtained at low concentrations to evaluate the substituent effects with the DSP approach. For most carbons, good fits to Eq. 4.3 were obtained, with the exception of the proximate carbons C-1, -2, and -9 in **71** and C-1, -2, and -3 in **72**. These sites are markedly affected by steric, neighboring group, magnetic anisotropy and possibly bond order effects that may overshadow the electronic phenomena.

An interesting aspect of their [85] work was the significant differences between the DSP correlative analysis of the ^{13}C shielding data and the DSP results for ^{19}F substituent effects and reactivity data obtained previously on a similar set of compounds.[87] As a good first approximation, polar and resonance effects were negligible at the 5α (C-5 in **71**) and 6α (C-5 in **72**) dispositions, as determined by the ^{13}C probe. The ^{19}F chemical shifts,[87] however, indicate substantial residual polar effects at both dispositions, with mesomerism approximately zero at the 5α disposition, but significant for the unconjugated 6α disposition. Thus, the nature of polar substituent effects probed by ^{13}C NMR and by ^{19}F NMR is quite different. The polar substituent effects at the 5α and 6α dispositions observed in the ^{19}F NMR spectra must arise from direct field effects—that is, from through-space interaction of the electrostatic field vector on the potential π component of the C-F bond.

Kitching and his colleagues[88] have extended their analyses to the substituent effects, $\Delta\delta_C$, in α-substituted 2-methylnaphthalenes, **86**, where X = H, SCH$_3$,

α-substituted 2-methylnaphthalenes
(**86**)

OCH$_3$, N(CH$_3$)$_2$, Br, Cl, or CN. Using values of ρ_I and ρ_R, the coefficients of the inductive (polar) and resonance terms of Eq. 4.3, derived from $\Delta\delta$ values for C-6, -7, and -10 of **72** in the earlier work,[86] σ_I and σ_R^0 values for the CH$_2$X groups were obtained from the $\Delta\delta$ values for these carbons in **86**. The σ_I and σ_R^0 thus derived were used to calculate the polar and resonance contributions to $\Delta\delta$, and to calculate $\Delta\delta$ itself for C-4 in a series of benzyl derivatives, **87**. The

α-substituted
toluenes
(**87**)

calculated and experimental $\Delta\delta$ values were in good agreement. Additionally, the coefficients, $\rho_I = 4.73$ and $\rho_R = 20.98$, of the DSP equation for C-4 in **87**,

$$\Delta\delta = 4.73\ \sigma_I + 20.98\ \sigma_R^0 \tag{4.4}$$

show the predominance of the resonance contribution; for all groups the resonance effect, $\rho_R\,\sigma_R^0$, was greater than the polar contribution, $\rho_I\sigma_I$. Good agreement was found with σ_R^0 ($CH_2 X$) values based on other techniques.

A paper by Sawhney and Boykin[89] treats the ^{13}C chemical shift data for 57 benzothiazoles—including 6-substituted 2-aminobenzothiazoles, **88**, 6-substituted 2-methylbenzothiazoles, **89**, 6-substituted benzothiazoles, **90**, 5-substituted 2-methylbenzothiazoles, **91**, and 2-substituted benzothiazoles, **92**—with

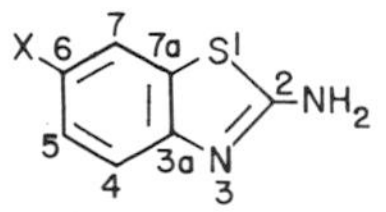

6-substituted
2-aminobenzothiazoles
(88)

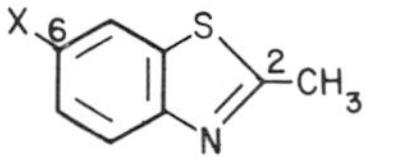

6-substituted
2-methylbenzothiazoles
(89)

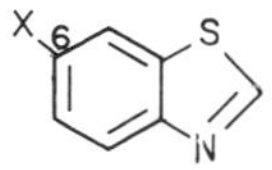

6-substituted
benzothiazoles
(90)

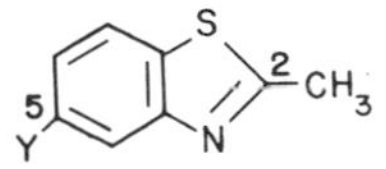

5-substituted
2-methylbenzothiazoles
(91)

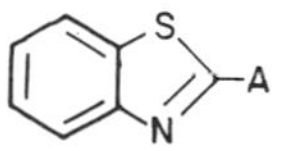

2-substituted
benzothiazoles
(92)

the Hammett, Swain–Lupton, and Taft DSP approaches. Results from the dual parameter approaches indicate the predominance of resonance effects in the determination of the ^{13}C substituent chemical shifts, just as we have seen previously for polycyclic aromatic hydrocarbons. Transmission of substituent effects by the sulfur atom in **88-92** appeared limited, with the primary path for this transmission to C-2 through the nitrogen.

The DSP analysis of substituted napthalenes by the Kitching group[86] mentioned earlier clearly indicates the dominance of the mesomeric contribution in these compounds. For several amino- and dimethylamino-naphthalenes, Ernst[90] showed that the effects of amino substituents on the ^{13}C shieldings of naphthalenes are transmitted through resonance to nearly all positions.

In rather large polycyclic aromatic molecules—pyrene, **5**, for example—the resonance transmission of substituent effects can be seen clearly.[91-94] Even for distant carbons, $\Delta\delta$ values alternate in magnitude and sign, depending on the mesomeric release or withdrawal of electrons. Thus, the effects of the amino substituent on the ^{13}C chemical shifts of 1-aminopyrene, **93**, are relatively large,

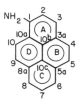

1-aminopyrene
(93)

even in ring C, which is most distant from the substituent. Carbon 6 in **93**, for example, is more shielded than C-6 in the parent compound pyrene, **5**, by 2.22 ppm ($\Delta\delta = -2.22$); for C-5a, $\Delta\delta = +1.18$.[93] The dominance of mesomeric transmission of the substituent effects is also indicated by the reduction of these effects that occurs when conjugation is inhibited for steric reasons,[90] as in 1-dimethylamino-2-methylnaphthalene, **94**.

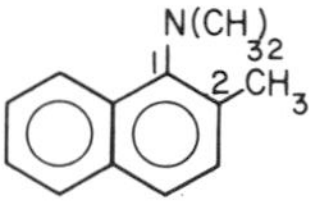

1-dimethylamino-
2-methylnaphthalene
(94)

A recent publication by Bromilow and his colleagues[95] reports that substituent effects in 1,4-disubstituted benzenes are not additive: a fixed group, Y, substantially changes the substituent chemical shifts at the para position. The Y group, through changes in the π-electron density at the para carbon, changes the sensitivity of the chemical shift of that carbon to an ipso substituent, and further produces a nonlinear dependence on the DSP resonance parameter. The data were analyzed in terms of a modification of the dual substituent parameter equation, which introduces an electron-demand parameter, ϵ, for the Y group, to account for the enhanced or reduced π-electron delocalization of the para group caused by Y, and involves a nonlinear relationship between $\Delta\delta$ and the σ_R^0 values of the para substituent:

$$\Delta\delta = \rho_I \sigma_I + \frac{\rho_R \, \sigma_R^0}{\epsilon(1 - \sigma_R^0)} \tag{4.5}$$

Use of this dual-substituent parameter with nonlinear regression equation, or DSP–NLR, produced fits that were significantly better than those from the DSP equation only when the coefficient of the resonance term, ρ_R, was large. Further, the DSP–NLR approach furnished the greatest improvement where either $\lambda = \rho_R/\rho_I$ was negative or large, or where the best DSP fit was with a σ_R other than the NMR-derived σ_R^0.

Wilson[82] extended this work to the analysis of substituent effects on the ^{13}C NMR spectra of 4-substituted paraterphenyls, **95**. For an extensive basis

4-substituted paraterphenyls
(95)

set of substituents—$R = NO_2, COOCH_3, CN, H, CH_3, I, Br, Cl, NH_2$, and $N(CH_3)_2$ — correlations between $\Delta\delta$ values and Hammett inductive and resonance σ parameters were sought, using three models. These models are the simple Hammett relationship given in Eq. 4.1, the Taft dual substituent relationship given in Eq. 4.3, and a more general form of the nonlinear dual substituent relationship given in Eq. 4.5. For convenience, these models are summarized here:

$$\Delta\delta = \rho\sigma \qquad \text{Hammett}$$

$$\Delta\delta = \rho_I\sigma_I + \rho_R\sigma_R \qquad \text{DSP}$$

$$\Delta\delta = \rho_I\sigma_I + \frac{\rho_R\sigma_R}{(1 - \epsilon\sigma_R)} \qquad \text{DSP–NLR}$$

Since the substituent effects measured were relatively small, stringent statistical tests were necessary to evaluate the data. No fit was considered acceptable at less than the 99% confidence level using a χ^2 test. The choice of the best model was then made using the F test for equality of variances, at the 95% confidence level.

For the substituted carbon, C-4, and for the carbons ortho to it, C-3 and C-5, no acceptable fits were obtained. The simple Hammett model was best at only two positions—C-2', -6' and C-2", -6". At most positions (C-1; C-2, -6; C-3', -5'; C-1"; C-3", -5"; and C-4") the DSP model and the resonance parameter σ_R^0 gave the best fit. The DSP–NLR model was best at the positions along the long molecular axis in the central ring, C-1' and C-4', reflecting the compensating changes in the π charge in this ring by redistribution and interaction with the terminal ring, in the presence of the substituent.

An interesting aspect of these results is that the ratio of resonance to inductive transmission of substituent effects, $\lambda = \rho_R/\rho_I$, had a rather small range, 0.54 to 4.6, indicating the importance of inductive electronic effects at all positions in the terphenyls, even at the carbon furthest from the substituent, C-4", where $\lambda = 1.7$.

Transmission of substituent effects on ^{13}C NMR chemical shifts in nonalternant π systems has been examined by a few authors. Bagli and St.-Jacques[96] analyzed the spectra of a series of 2-substituted tropones, **96**, which are nonbenzenoid aromatic substances. Substituent effects, $\Delta\delta$, were obtained for the

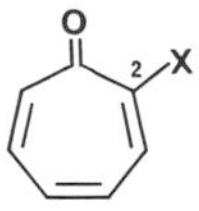

2-substituted
tropones
(96)

substituents X = OH, OCH_3, $OCOCH_3$, NH_2, $NHCH_3$, $NHCOCH_3$, Cl, CH_3, and SCH_3. These effects were then compared to the corresponding $\Delta\delta$ values for the same substituents in benzenes. With the exception of the SCH_3-substituted tropones, linear relationships were obtained between the tropone and the benzene values. Thus π-electron redistribution through mesomeric effects is important in tropones as well as in benzenes. However, the linear relationships divided the substituents into two classes, with the origin of the differences between these classes not easily explainable.

A recent investigation[97] of transmission of electronic substituent effects in the nonalternant 1- and 2-substituted azulenes, **97** and **98**, employed ^{13}C NMR

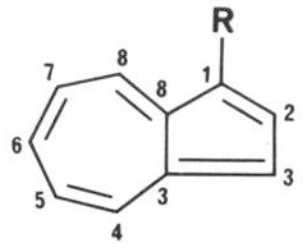

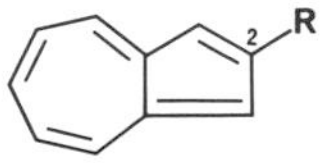

1-substituted   2-substituted
azulenes   azulenes
(97)   (98)

chemical shifts and CNDO molecular orbital calculations of electron densities. The data were analyzed with the DSP model, to separate the resonance and inductive substituent components, then compared with the results of the molecular orbital calculations. These comparisons showed good agreement between the CNDO-derived charge densities and the electron distributions implied by the coefficients obtained in the DSP analyses. In contrast to alternant aromatic compounds such as naphthalenes, the greatest sensitivity of ^{13}C chemical shifts to the substituent in these nonalternant systems occurs in the seven-membered ring, remote from the site of substitution. Both field and resonance effects contribute significantly to the observed variations in $\Delta\delta$. In the five-membered ring these effects oppose each other, whereas in the seven-membered ring they enhance each other. Hence the differences in sensitivity between the two rings.

Direct Electric Field Effects

Earlier in this chapter we discussed the differences between the ^{13}C and ^{19}F substituent effects in naphthalenes.[86] Although field-induced π-charge polarization and mesomerism were negligible at certain positions relative to the substituents, the $\Delta\delta_F$ values observed in the ^{19}F NMR spectra for these positions were substantial. This result was attributed to a large contribution to the ^{19}F shielding from the direct influence of the electric field associated with the substituent on the C–F bond.

Evidence for direct electric field effects on ^{13}C NMR shieldings is scant. In a series of naphthalenes with electron-withdrawing substituents, **71** and **72** (R = NO_2, COOH, CN, CHO, and $COCH_3$), Ernst[67] found that most of the $\Delta\delta$ values at positions ortho to R were positive—that is, the substituents decreased the shielding at the ortho position, as would be expected for a dominant mesomeric contribution. However, certain of these substituent effects were negative, resulting from increased shielding of the ortho ^{13}C nuclei—for example, $\Delta\delta = -1.98$ for C-2 in 1-nitronaphthalene, **99**. This shielding effect of the nitro substituent

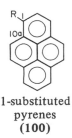

1-nitronaphthalene
(99)

on the ortho carbon was ascribed primarily to polarization of the electrons of the C_{ortho}–H bond by the electrostatic field of the nitro group, decreasing the electron density at the proton and increasing it at the carbon nucleus. The peri carbon, C-8, in these compounds, **71**, is also considerably shielded—for example, $\Delta\delta = -5.23$ for R = NO_2 and $\Delta\delta = -3.24$ for R = CN. Although part of this effect may result from sterically induced polarization of the C–H bond,[61] it is conceivable that direct electric field effects play an important part, particularly for substituents that are not bulky.

Hansen and his co-workers[92,94] have suggested that the high-field shifts of C-10a in pyrene carboxylic acid derivatives of hydroxy-substituted pyrenes, **100**,

1-substituted
pyrenes
(100)

(R = COOR' or OH), may be caused by electric field effects. Further investigation of these effects is certainly warranted, but separation of the influences of direct electric fields and π-charge polarization may be difficult.

Additivity

The Use of Additivity in Simple Molecules

There have been many examinations of the proposal[98,99] that substituent effects on ^{13}C shieldings of aromatic compounds are approximately additive. These examinations have usually shown that, if compounds having significant steric interactions are excluded, ^{13}C substituent effects are indeed additive. The finding has proved to be of great utility in the assignment of spectra and the analysis and characterization of new compounds.

The basis of prediction of chemical shifts using empirical additive substituent parameters is the selection of suitable model compounds and the use of a number of substituted compounds sufficient for regression analysis. The effectiveness of this approach was first demonstrated by Grant and Paul[100] for paraffinic hydrocarbons. The method was later extended to simple aromatic compounds.[1-4] To predict the ^{13}C chemical shifts, $\delta_C(i)$, in a substituted benzene, for example, the appropriate substituent parameters, $\Delta\delta_i$, from Table 4.8 are added to the chemical shift of benzene itself:

$$\delta_C(i) = 128.5 + \sum_i \Delta\delta_i(R) \qquad (4.6)$$

Similar parameters (Table 4.9) have been developed for substituted pyridines:

$$\delta_C(i) = \delta_C^0(i) + \sum_i \Delta\delta_i(R) \qquad (4.7)$$

Here $\delta_C(i)$ is the ^{13}C chemical shift of carbon i in the substituted compound and $\delta_C^0(i)$ is the chemical shift of this carbon in pyridine. Thus, the predicted ^{13}C chemical shifts of the aryl carbons in 2,6-dimethylpyridine are $\delta_C(2,6) = 158.6$, $\delta_C(3,5) = 119.8$, and $\delta_C(4) = 136.0$. These values compare favorably with the experimental values $\delta_{C-2,-6}$ 158.1, $\delta_{C-3,-5}$ 120.7, and δ_{C-4} 137.7.[1,102] It is possible to predict the δ_C values for benzenes or pyridines having several different substituents by using this simple additivity approach. Significant complications can arise when steric interactions or other interactions such as hydrogen bonding occur. Deviations from the additivity predictions result. These deviations are discussed later in this chapter.

The Use of Additivity in Complex Molecules

One of the most fortunate and useful properties of additive substituent parameters for prediction of ^{13}C chemical shifts is their transferability from one structurally similar molecular fragment to another. Initial assignments of ^{13}C resonances in substituted naphthalenes and biphenyls could thus be made on the basis of the effects of similar substituents in benzenes. Likewise, substituent

Table 4.8. Substituent Parameters for the Calculation of ^{13}C Chemical Shifts^a in Substituted Benzenes

R	Ipso	Ortho	Meta	Para
		$\Delta\delta_i$		
H	0	0	0	0
CH_3	+ 9.3	+ 0.8	0	− 2.9
CH_2CH_3	+15.6	− 0.4	0	− 2.6
$CH(CH_3)_2$	+20.2	− 2.5	+0.1	− 2.4
$C(CH_3)_3$	+22.4	− 3.1	−0.1	− 2.9
CF_3	− 9.0	− 2.2	+0.3	+ 3.2
C_6H_5	+13	− 1	+0.4	− 1
$CH{=}CH_2$	+ 9.5	− 2.0	+0.2	− 0.5
$C{\equiv}CH$	− 6.1	+ 3.8	+0.4	− 0.2
CH_2OH	+12	− 1	0	− 1
COOH	+ 2.1	+ 1.5	0	+ 5.1
COO^-	+ 8	+ 1	0	+ 3
$COOCH_3$	+ 2.1	+ 1.1	+0.1	+ 4.5
COCl	+ 5	+ 3	+1	+ 7
CHO	+ 8.6	+ 1.3	+0.6	+ 5.5
$COCH_3$	+ 9.1	+ 0.1	0	+ 4.2
$COCF_3$	− 5.6	+ 1.8	+0.7	+ 6.7
COC_6H_5	+ 9.4	+ 1.7	−0.2	+ 3.6
CN	−15.4	+ 3.6	+0.6	+ 3.9
OH	+26.9	−12.7	+1.4	− 7.3
OCH_3	+31.4	−14.4	+1.0	− 7.7
$OCOCH_3$	+23	− 6	+1	− 2
OC_6H_5	+29	− 9	+2	− 5
NH_2	+18.0	−13.3	+0.9	− 9.8
$N(CH_3)_2$	+23	−16	+1	−12
$N(C_6H_5)_2$	+19	− 4	+1	− 6
$NHCOCH_3$	+11	−10	0	− 6
NO_2	+20.0	− 4.8	+0.9	+ 5.8
NCO	+ 5.7	− 3.6	+1.2	− 2.8
F	+34.8	−12.9	+1.4	− 4.5
Cl	+ 6.2	+ 0.4	+1.3	− 1.9
Br	− 5.5	+ 3.4	+1.7	− 1.6
I	−32	+10	+3	+ 1

Source. References 1–4.

^{a}The ^{13}C chemical shift of unsubstituted benzene is δ_C 128.5.

Table 4.9. Substituent Parameters for the Calculation of Chemical Shifts[a] of Substituted Pyridines

| 2-R | $\Delta\delta_i$ | | | | |
| | i | | | | |
	2	3	4	5	6
CH_3	+ 9.1	− 1.0	− 0.1	− 3.4	− 0.1
CH_2CH_3	+14.0	− 2.1	+ 0.1	− 3.1	+ 0.2
$COCH_3$	+ 4.3	− 2.8	+ 0.7	+ 3.0	− 0.2
CH	+ 3.5	− 2.6	+ 1.3	+ 4.1	+ 0.7
OH	+14.9	−17.2	+ 0.4	− 3.1	− 6.8
OCH_3	+15.3	−13.1	+ 2.1	− 7.5	− 2.2
NH_2	+11.3	−14.7	+ 2.3	−10.6	− 0.9
NO_2	+ 8.0	− 5.1	+ 5.5	+ 6.6	+ 0.4
CN	− 15.8	+ 5.0	− 1.7	+ 3.6	+ 1.9
F	+14.4	−14.7	+ 5.1	− 2.7	− 1.7
Cl	+ 2.3	+ 0.7	+ 3.3	− 1.2	+ 0.6
Br	− 6.7	+ 4.8	+ 3.3	− 0.5	+ 1.4

| 3-R | i | | | | |
	2	3	4	5	6
CH_3	+ 1.3	+ 9.0	+ 0.2	− 0.8	− 2.3
CH_2CH_3	+ 0.3	+15.0	− 1.5	− 0.3	− 1.8
$COCH_3$	+ 0.5	− 0.3	− 3.7	− 2.7	+ 4.2
CHO	+ 2.4	+ 7.9	0	+ 0.6	+ 5.4
OH	− 10.7	+31.4	− 12.2	+ 1.3	− 8.6
NH_2	− 11.9	+21.5	− 14.2	+ 0.9	−10.8
CN	+ 3.6	− 13.7	+ 4.4	+ 0.6	+ 4.2
Cl	− 0.3	+ 8.2	− 0.2	+ 0.7	− 1.4
Br	+ 2.1	− 2.6	+ 2.9	+ 1.2	− 0.9
I	+ 7.1	− 28.4	+ 9.1	+ 2.4	+ 0.3

| 4-R | i | | |
	2	3	4
CH_3	+ 0.5	+ 0.8	+10.8
CH_2CH_3	0	− 0.3	+15.9
$CH{=}CH_2$	+ 0.3	− 2.9	+ 8.6
$COCH_3$	+ 1.6	− 2.6	+ 6.8
CHO	+ 1.7	− 0.6	+ 5.5
NH_2	+ 0.9	− 13.8	+19.6
CN	+ 2.1	+ 2.2	− 15.7
Br	+ 3.0	+ 3.4	− 3.0

Source. References 1–4, 10, and 101.
[a]The ^{13}C chemical shifts of unsubstituted pyridine are $\delta_C^0(2,6) = 149.6$, $\delta_C^0(3,5) = 124.2$, and $\delta_C^0(4) = 136.2$.

effects in pyrene[91,93] derivatives have been predicted with some success, using averaged substituent parameters from naphthalenes.

Values of $\Delta\delta_C$ for chloro, nitro, and methyl substituents in similar positions in several polycyclic aromatic hydrocarbons are given in Table 4.10. Agreement between $\Delta\delta$ for a given substituent in benzene and in the other compounds is good for the substituted (ipso) and para carbons, reflecting the predominance of directly bonded and mesomeric transmission of the substituent effects. At the ortho positions, agreement is lessened by the steric and possible direct electric field effects (in the case of the nitro) of the substituent. At the meta positions, the effects of the substituent are relatively small and experimental error becomes a significant factor.

Although initial assignments can be made by using additivity parameters derived for another similar molecule, the accuracy of the additivity predictions clearly depends on the validity of the model compound chosen. With a good choice of model compound, it is possible to predict the ^{13}C chemical shifts within less than ±1 ppm.

When substituent parameters are available for the parent compound, better precision can be obtained. For example, using the 1-CH$_3$ and 2-CH$_3$ substituent parameters given in Table 4.10 for naphthalene, the ^{13}C chemical shifts of all the mono- and dimethylnaphthalenes, excluding those with ortho or peri methyls, are predicted within ±0.15 ppm.[13] Similarly, the chemical shifts of 67 independent aryl carbons in methylanthracenes lacking peri methyls are predicted within ±0.12 ppm,[14] as is illustrated in Figure 4.9. Similar results have been obtained for methylated azulenes.[68]

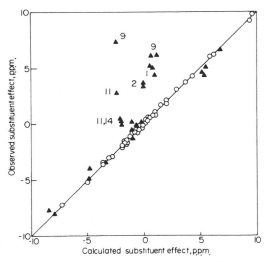

Figure 4.9. Observed versus calculated methyl substituent effects on aryl ^{13}C chemical shifts in some methylated anthracenes. (From reference 14.)

Table 4.10. Comparisons of Some Substituent Effects on the ^{13}C Chemical Shifts of Benzene, Naphthalene, Anthracene, Biphenyl, and Terphenyl

Substituent	Parent Compound	Ipso	Ortho	Meta	Para	Reference
				$\Delta\delta$, ppm		
1-Cl	Benzene	+ 6.2	+0.4	+1.3	-1.9	1-4
1-Cl	Naphthalene	+ 4.0	+0.3 (C-2)	+0.0 (C-3)	-0.9	103
			-2.6 (C-8a)	+1.2 (C-4a)		
2-Cl	Naphthalene	+ 5.9	+1.0 (C-3)	+1.8 (C-4)	-1.8	103
			-1.2 (C-1)	+0.7 (C-8a)		
2-Cl	Biphenyl	+ 5.0	+2.4 (C-3)	+1.1 (C-4)	-2.1	60
			-2.1 (C-1)	+2.6 (C-6)		
3-Cl	Biphenyl	+ 5.5	+0.6 (C-2)	+1.6 (C-1)	-2.0	60
			-0.2 (C-4)	+1.1 (C-5)		
4-Cl	Biphenyl	+ 5.9	+0.0	+1.1	-1.7	60
4-Cl	p-Terphenyl	+ 6.0	+0.0	+1.3	-1.5	104
1-CH$_3$	Benzene	+ 9.3	+0.8	0.0	-2.9	1-4
1-CH$_3$	Naphthalene	+ 6.2	+0.8 (C-2)	-0.3 (C-3)	-1.6	13
			-0.8 (C-8a)	+0.2 (C-4a)		
2-CH$_3$	Naphthalene	+ 9.6	-1.0 (C-1)	-0.6 (C-4)	-1.8	13
			+2.3 (C-3)	+0.2 (C-8a)		
1-CH$_3$	Anthracene	+ 5.8	+0.3 (C-2)	-0.2 (C-3)	-1.5	14
			-0.4 (C-9a)	0.0 (C-4a)		
2-CH$_3$	Anthracene	+ 9.6	-1.5 (C-1)	0.0 (C-4)	-1.2	14
			+2.9 (C-3)	+0.4 (C-9a)		
9-CH$_3$	Anthracene	+ 3.7	-1.6	-0.3 (C-4a)	-1.7	14
4-CH$_3$	Biphenyl	+ 9.6	+0.8	-0.1	-2.8	69
1-NO$_2$	Benzene	+20.0	-4.8	+0.9	+5.8	1-4
1-NO$_2$	Naphthalene	+19.4	-1.5 (C-2)	-1.3 (C-3)	+7.3	105
			-8.1 (C-8a)	+1.3 (C-4a)		
4-NO$_2$	Biphenyl	+13.3	-3.8	+1.4	+3.8	69
4-NO$_2$	p-Terphenyl	+19.8	-4.6	+0.8	+6.4	81, 82

Thus, using additive substituent parameters, ordering of the resonance assignments can be accomplished, but unequivocal assignments must be based on other techniques, such as off-resonance and selective single frequency decoupling of the protons, selective deuteration, relaxation time measurements, or lanthanide-induced shift determinations. These assignment techniques will be discussed in Chapter 7.

The utility of using additivity to predict ^{13}C NMR chemical shifts is demonstrated by the identification[103] of an unknown dichloronaphthol metabolite of 2,6-dichloronaphthalene, **101**. The mass and ^{1}H NMR spectra showed that the

2,6-dichloronaphthalene
(101)

metabolite was a dichloronaphthol,[104] but could not differentiate unambiguously between 2,6-dichloro-1-hydroxynaphthalene, **102**, and 1,6-dichloro-2-hydroxynaphthalene, **103**. Additivity predictions were used to order the resonances of the metabolite. After these were grouped into protonated carbon

2,6-dichloro-1-
hydroxynaphthalene
(**102**)

1,6-dichloro-2-
hydroxynaphthalene
(**103**)

resonances and quaternary carbon resonances, but without specific assignments, the differences between the predicted and the observed chemical shifts were calculated. For structure **102**, the average difference was 1.95 ppm, but for structure **103**, it was only 0.47 ppm. The results clearly established the metabolite structure as the 2-naphthol **103**.[103]

Of the substituted polycyclic aromatic hydrocarbons, naphthalene is the parent compound for which the most ^{13}C NMR data are available in the literature (references 13, 28, 67, 86, 88, 90, 103, 105, 106–114). Hansen has collected this vast amount of data in his review article,[6] and the reader is referred to this source for more information on the ^{13}C chemical shifts of substituted naphthalenes.

A smaller number of studies has concerned substituent effects on the ^{13}C shieldings of phenanthrenes,[16] fluorenes,[16,84,115] pyrenes,[15,93] fluoranthenes,[116] anthracenes,[14] azulenes,[68] indoles and indenes,[117,118] quinolines and isoquinolines,[32,28,33,119-121] phenazines,[34] purines,[38,39,122] biphenyls,[60,69,123] terphenyls,[81,82,124] and bipyridyls and phenanthrolines.[43]

Steric Effects and Deviations From Additivity

Steric effects on ^{13}C NMR chemical shifts resulting from the introduction of substituents have been observed in a number of systems. For an extensive review of this topic, the reader is referred to the work of Wilson and Stothers.[59]

Grant and Cheney[61] sought to explain the increased shieldings of ortho methyl carbons in methyl-substituted benzenes by use of a model invoking steric polarization of the C–H valence electrons by the nonbonded repulsions of the sterically interacting hydrogen atoms so that the charge densities at the carbons are changed. The changed charge densities are reflected in the effective nuclear charges, which then lead to changes in the dominant paramagnetic contributions to the ^{13}C chemical shifts. Thus, in an ortho dimethyl substituted naphthalene—for example, 2,3-dimethylnaphthalene, **104**—the steric polarization of the C–H

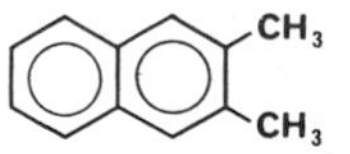

2,3-dimethylnaphthalene
(104)

bonds in the interacting methyls would lead to increased charge densities at the methyl carbon nuclei; hence increased shielding or an upfield shift is observed.

This bond polarization argument has been used to explain steric effects on ^{13}C shieldings in a number of alkanes and alkenes. Grant and Cheney's analysis was used by Clark[125] to calculate the δ values for the methyl carbon and C-8 in 1-methylnaphthalene, **58**, and some methylbenzothiophenes, but with only qualitative agreement.

For neighboring methyl carbon atoms separated by three bonds, such as the methyl carbons in *cis*-2-butene, **105**, relative to those in the trans isomer, **106**,

cis-2-butene   trans-2-butene
(105)   **(106)**

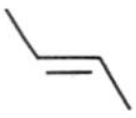

and ortho methyl carbons relative to para methyl carbons in aromatic compounds, the crowded methyl carbons are more shielded. With four intervening bonds, as is the case for the methyl carbons in 1,8-dimethylnaphthalene, **62**, or the 9-methyl carbon in 1,8,9-trimethylanthracene, **107**, decreased shielding is

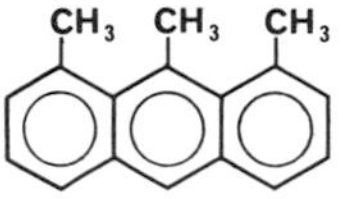

1,8,9-trimethylanthracene
(107)

apparent. In a series of 4,5-dimethyl-substituted phenanthrenes, fluorenes, and fluorenones,[16] the methyl carbons are five bonds apart, and the methyl groups are more shielded than predicted. However, closer scrutiny of the effects of the dimethyl substitution on the aromatic skeletons shows that there is no consistent pattern and that explanations based on numbers of intervening bonds and bond polarization models alone are too simple.

Since mesomeric transmission of substituent influences plays a major role in substituent effects on ^{13}C chemical shifts of aromatic carbons, steric interactions that change the effectiveness of resonance in the molecule will change the substituent effects. In the N-methylaminonaphthalenes discussed earlier, the Δδ values were smaller than anticipated, presumably because of the steric inhibition of resonance.[90]

In addition to bond polarization and resonance reduction, distortions of the molecular geometry through sterically produced bond angle changes probably play an important part in ^{13}C substituent effects. For example, in acenaphthene, **21**, the ortho and para carbons are much more shielded, δ_{C-3} 118.9 and δ_{C-5} 122.0,[126] than the analogous carbons in 1,8-dimethylnaphthalene, **62**, δ_{C-2} 129.2 and δ_{C-4} 127.7.[13] This result has been attributed[126] to ring strain in **21**.

A similar comparison of the substituted carbon chemical shifts in 1,8-dichloronaphthalene,[103] **108**, δ_{C-1} 130.43 and δ_{C-8a} 127.46, with their counterparts in

1,8-dichloronaphthalene
(108)

62, δ_{C-1} 135.2 and δ_{C-8a} 132.9, shows greater shielding of these carbons in the chlorine-substituted compound. Ernst[28,112] has suggested that the discrepancy results from greater distortion of the substituted part of the molecule and greater opening of the R–C-1–C-9 angle in **108** than in **62**.

Thus, we can see that the well-documented successes of application of the additivity principle to prediction of ^{13}C NMR chemical shifts should not be used to justify assignments based on additivity in all cases. One critical factor in the successes of additivity predictions is the choice of suitable model compounds and substituent parameters, as we mentioned earlier. Another critical factor is the absence of significant steric interactions.

In compounds in which steric compression of the substituents occurs, the deviations from additivity are fairly large, as can be seen from examination of the data given in Table 4.11 for several substituted naphthalenes. Ortho disubsti-

Table 4.11. Deviations from Additivity in Some Sterically Hindered Substituted Naphthalenes

| Substitution | $\Delta\delta$, ppm | | | | | | | | | | Reference |
	C-1	C-2	C-3	C-4	C-5	C-6	C-7	C-8	C-4a	C-8a	
1,2-(CH$_3$)$_2$	-2.1	-3.2	+1.2	+0.1	+0.2	-0.2	-0.0	-0.1	+0.5	0.0	13
2,3-(CH$_3$)$_2$	+1.2	-2.3	-2.3	+1.2	-0.5	-0.1	-0.1	-0.5	+0.5	+0.5	13
1,8-(CH$_3$)$_2$	+5.1	+3.0	-0.2	+1.0	+1.0	-0.2	+3.0	+5.1	+1.7	+1.2	13
1,2-Cl$_2$	-1.1	-0.1	+0.0	-0.5	-0.1	+0.0	-0.3	+0.8	-0.1	-1.0	103
2,3-Cl$_2$	-1.2	-2.4	-2.4	-1.2	+0.0	+1.4	+1.4	+0.0	+1.1	+1.1	103
1,8-Cl$_2$	+1.8	+3.8	-0.7	+1.4	+1.4	-0.7	+3.8	+1.8	+1.5	-0.7	103
1-Cl,2-OH	-0.1	-4.1	-0.4	-0.7	+0.5	-0.5	+0.1	-0.4	-0.5	-0.8	103
2-Cl,1-OH	-2.7	-1.0	-0.5	-1.5	-0.0	-0.1	-0.6	+1.7	+0.5	-0.3	103
1-Cl,8-OH	+1.9	+1.8	-0.9	+1.6	-0.2	+0.6	+3.5	+4.8	+1.3	-1.8	103

tution, as in 1,2- and 2,3-dimethylnaphthalene, usually leads to marked increases in the shieldings of the substituted carbons as well as the methyl carbons, and concomitant decreases in the shieldings of the carbons nearest the substituted carbons. These shifts of the substituted carbon resonances to higher fields are precisely as expected from the Grant–Cheney bond polarization model[61] and the well-established observation that crowded alicyclic carbons separated by three bonds absorb at higher fields than their less crowded counterparts.[59]

Similarly, the decreased shielding of the carbons in the crowded portion of the molecule in the peri-disubstituted naphthalenes can be *partly* rationalized with the bond polarization model. A simple charge density dependence does not suffice to explain the observed deviations, as shown by the modest correlations of charge densities calculated by the CNDO/2 and INDO methods[13,67] with the observed ^{13}C chemical shifts in these compounds. Thus, contributions from other factors, including bond angle changes,[28] may also be important.

Clearly, the effects of sterically caused changes in bond polarizations and charge densities, bond angles, and resonance cannot always be separated. For example, in chlorinated biphenyls, ortho–ortho' disubstitution can cause changes in the minimum energy inter-ring dihedral angle.[60] This angle, in 2,6,2',6'-tetra-chlorobiphenyl, **75**, is close to 90°, whereas in biphenyl **3** itself, this angle is about 43°. Thus, delocalization of the π-electron system is markedly reduced in **75** relative to **3**. Careful consideration of the C-1-C-2-Cl bond angles in **75** indicates that these angles are probably greater than the corresponding C-C-H angles. CNDO/2 calculations show marked changes in the carbon charge densities[69] in **75** relative to **3**. The changes in the effects of the ring currents in one ring on the nuclei in the other ring with changes in the inter-ring dihedral angle may also contribute to substituent effects in biphenyls and similar systems.

All of these factors lead to large deviations from additivity predictions in the chlorinated biphenyls. Similar results can be expected, of course, for other sterically hindered compounds.

CHEMICAL SHIFTS OF HETEROATOMS IN AROMATIC COMPOUNDS

Several of the heteroatoms that can become part of the skeleton of aromatic compounds, such as nitrogen, oxygen, sulfur, and selenium, have magnetic isotopes. Unfortunately, most of these magnetic isotopes present some difficulties to the NMR spectroscopist. These difficulties may arise from low natural abundances, as for ^{15}N, ^{17}O, ^{33}S, and ^{77}Se, whose natural abundances are 0.365%, 0.037%, 0.74%, and 7.5%, respectively. The difficulties may arise from nuclear magnetogyric ratios that lead to small or negative nuclear Overhauser enhancements, as for ^{15}N and ^{17}O. Additional difficulties may arise from the quadrupole moments of nuclei with spin $I > 1/2$, which produce excessive line broadening and very short relaxation times, as for ^{14}N, ^{17}O, and ^{33}S, which have spin $I = 1$, $5/2$, and $3/2$, respectively.

Nevertheless, the effort required to examine the NMR spectra of these heteronuclei is often worthwhile. Fundamental information on the bonding, electronic structures, and reactions of aromatic heterocycles can often be gathered, and insights into the nature of the biological interactions in which these compounds are involved can often be obtained from the NMR data. Nitrogen-15 NMR is of particular importance in the latter regard.

Nitrogen-15

Because of the important role played by nitrogen in organic, biological, and inorganic chemistry, the potential value of nitrogen NMR is huge. Nitrogen-14, the more abundant isotope, has spin $I = 1$, so its resonance lines are generally broad, which limits the resolution obtainable.[127] Thus ^{15}N, despite the low NMR sensitivity and abundance of this spin-1/2 nucleus, has become important. Adequate instrumental techniques and a fundamental base of information about ^{15}N NMR spectroscopy have now been developed, so that ^{15}N NMR experiments on natural abundance samples, while still not routine, are practical. A recent text by Levy and Lichter[128] summarizes developments in ^{15}N NMR spectroscopy and its applications through early 1979.

Although nitrogen is similar to carbon in its structural, bonding, and electronic characteristics in aromatic compounds, there are several significant differences in the NMR characteristics of ^{15}N and ^{13}C. First, the nuclear Overhauser enhancement, which is often at the maximum for ^{13}C nuclei because of dipolar interactions with neighboring protons, is quite variable for ^{15}N nuclei, which frequently do not have dipole–dipole interactions dominating their spin–lattice relaxation. In some cases, this variability means that the NOE will cause a *decrease* in signal intensity, even to zero. Second, because of this insufficiency of dipolar interactions, the spin–lattice relaxation times may be very long, even for ^{15}N nuclei with nearby protons. Third, because nitrogen atoms are so often involved in tautomerism and other low-frequency processes, the spin–spin relaxation times may be short, leading to line broadening.

The range of ^{15}N chemical shifts is somewhat greater than that of ^{13}C, about 500 ppm compared to about 200 ppm. Nitrogen nuclei in molecular environments similar to those of carbon are subject to similar influences; hence, ^{15}N chemical shifts tend to parallel ^{13}C chemical shifts. Because of the unshared electron pair on nitrogen, ^{15}N chemical shifts are more sensitive to intermolecular interactions and solvent effects than are ^{13}C chemical shifts. A marked change in the ^{15}N resonance position is, of course, expected on protonation or substitution of the nitrogen. Although additive substituent parameters have been derived for ^{15}N chemical shifts in some aliphatic amines[129] and ureas,[130] these correlations have not been widely sought, and are not yet established for nitrogen in heterocyclic aromatics.

In general, a nitrogen in an aromatic ring is in one of two environments. In the chemical environment typified by pyrrole, **109**, the unshared electron pair

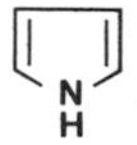

pyrrole
(109)

participates in the π-electron circulation of the aromatic ring. The nitrogen in pyrrole thus behaves much like a nitrogen in a typical aromatic amine such as aniline, **110**. The ^{15}N nuclei in pyrrole-type compounds tend to be more

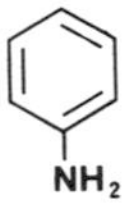

aniline
(110)

shielded—that is, their resonances are at higher field—than those in pyridine-type compounds. In the pyridine, **111**, type of environment, the nitrogen lone pair,

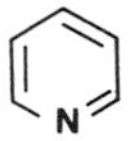

pyridine
(111)

which is orthogonal to the ring π system, is not incorporated into the π-electron circulation. Instead, it contributes an additional low-lying n–π^* state to the molecule, which decreases the mean excitation energy, ΔE, and hence decreases the nitrogen shielding. The ^{15}N chemical shift difference between these two molecular environments may be as great as 100 ppm. Some representative ^{15}N chemical shifts are given in Table 4.12. For more extensive summaries, the reader should refer to Levy and Lichter.[128]

Oxygen-17, Sulfur-33, and Selenium-77

Oxygen-17 nuclear magnetic resonance, although not yet extensively applied, has the potential to provide new insights into molecular structure. It has been used in several instances for the examination of such systems as polyoxomet-

Table 4.12. Some Representative ^{15}N NMR Chemical Shifts[a]

Compound	Solvent	δ_N	Reference
Pyrrole, **109**	$CDCl_3$	145.6	131
Pyridine, **111**	neat	317.4	135
	C_6D_6	320.7	135
Indole	$CDCl_3$	124.8	128
	$(CH_3)_2SO$	132.9	

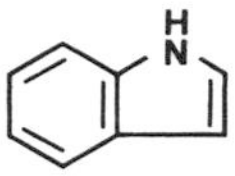

(112)

Pyrazole	$CHCl_3$	247.3 (N-1,2)	132
	CH_3OH	200.0 (N-1,2)	132

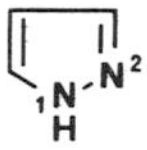

(113)

N-methylimidazole	CH_3OH	163.3 (N-1)	133
		248.0 (N-3)	
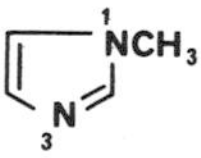	C_6H_{12}	160.1 (N-1)	133
		264.4 (N-3)	
	neat	161.8 (N-1)	134

(114)

N-methylpyrazole	$CHCl_3$	199.4 (N-1)	132
	$CDCl_3$	301.7 (N-2)	132

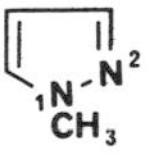

(115)

Quinoline	CCl_4	313.4	136

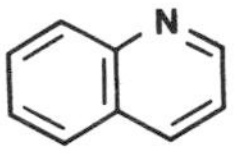

(116)

Pyrimidine	neat	294.8	134

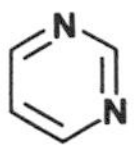

(117)

Quinazoline	$(CH_3)_2SO$	283.6 (N-1)	137
		295.0 (N-3)	

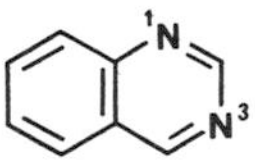

(118)

Thiazole	neat	323.0	134

(119)

Isothiazole		298.4	128

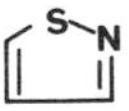

(120)

Table 4.12. (Continued)

Compound	Solvent	δ_N	Reference
Isoxazole (121)	neat	380.2	134
Dimethylfurazan (122)	$(CH_3)_2CO$	406.8	138
Benzofurazan (123)	$(CH_3)_2CO$	418.3	138
2,1,3-Benzothiadiazole (124)	$(CH_3)_2CO$	332.4	138
2,1,3-Benzoselenadiazole (125)	$(CH_3)_2CO$	375.0	138
Purine (126)	$(CH_3)_2SO$	281.3 (N-1) 262.2 (N-3) ~188.8 (N-7,9)	137
Indazolium ion (127)	aqueous HCl	206.1 (N-1) 174.2 (N-2)	139
Pyrazolium ion (128)	aqueous HCl	178.8	132
N-methylpyrazolium ion (129)	aqueous HCl	183.8 (N-1) 171.7 (N-2)	132
N,N-dimethylpyrazolium ion (130)	aqueous HCl	179.5	132

[a] Relative to anhydrous liquid ammonia at 25°C. Conversion factors were taken from Levy and Lichter,[128] pp. 32–33.

alates,[140] amino acids and model peptides,[141,142] and aliphatic ethers.[143] These applications, up to 1978, have been reviewed by Klemperer.[144]

The oxygen chemical shift range spans about 1000 ppm, but for oxygen atoms bonded only to carbon and/or hydrogen the range is smaller—fewer than 700 ppm. There is a clear correlation between the C-O bond order and the ^{17}O chemical shift. The more π character in the bridging oxygen C-O bond, the less shielded the ^{17}O nucleus. Bridging oxygens in the few heterocycles examined by ^{17}O NMR have resonances near the middle of the shift range. For example, the data for furan in Table 4.13 show the effect of π-electron delocalization, which shifts the bridging ^{17}O resonances downfield relative to the normal ether range (0–100 ppm). Bonding to nitrogen, as in the furazans **133** and **134** listed in the table, further decreases the ^{17}O shielding by π-electron delocalization. Although ^{17}O NMR experiments can be difficult because of the low natural abundance, negative magnetogyric ratio, and spin 5/2 of this nucleus, with current pulsed Fourier transform NMR instrumentation, ^{17}O data are relatively accessible. The same statement about accessibility cannot be made for sulfur-33.

The first investigations of ^{33}S NMR in aromatic compounds and its potential

Table 4.13. ^{17}O NMR Chemical Shifts[a] in Some Heterocycles

Compound	δ_O	
	Bridging	Terminal
Furan  (131)	241	
2-Formylfuran 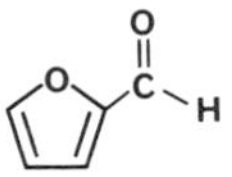 (132)	237	530
3,4-Dimethylfurazan 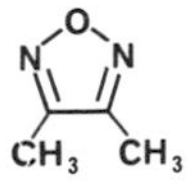 (133)	460	
3,4-Dimethylfurazan *N*-oxide 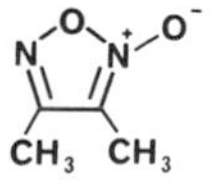 (134)	350	475

[a]Relative to H_2O. Data from reference 145.

were done by Retcofsky and Friedel[146,147] in the early 1970s. The ^{33}S spectrum of thiophene, **46**, exemplifies the difficulties inherent in observation of NMR signals of this nucleus. The sulfur resonance in **46** was found, after some difficulty,[146] to be 231 ppm from the carbon disulfide resonance; its line width at half height was 710 Hz, which at 4.33 MHz is approximately 164 ppm. The authors could not observe the ^{33}S resonance in the nonaromatic analog of **46**, tetrahydrothiophene, **135**, at all. In their later work,[147] Retcofsky and Friedel

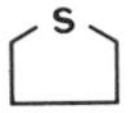

tetrahydrothiophene
(135)

were able to observe resonances of ^{33}S nuclei in several thiophene derivatives, and in some inorganic sulfur compounds. Some of these data are given in Table 4.14. In addition to chemical shifts, some sulfur relaxation times were measured and found to be on the order of milliseconds. Due to the asymmetry of the ^{33}S nuclear charge distribution, typical ^{33}S line widths are expected to be about 700 times those of ^{17}O. In general, ^{33}S chemical shifts can be expected to be of low accuracy, and the probability of differentiating by ^{33}S NMR between sulfur types in various environments is low. The demands on instrumentation resulting

Table 4.14. ^{33}S NMR Chemical Shifts[a] of Some Thiophenes

Compound	δ_S
Thiophene, **46**	220
3-Bromothiophene	134
(136)	
2-Methylthiophene	178
(137)	
3-Methylthiophene	197
(138)	

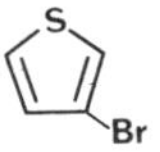

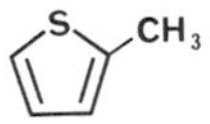

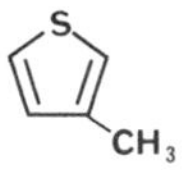

[a]Relative to $\delta_{CS_2} = 0$. Samples were 90% solutions in CS_2. Data from reference 147.

Table 4.15. ^{77}Se NMR Chemical Shifts[a] of Some Selenophenes

Compound	δ_{Se}	Reference
Selenophene, **47**	0.0	152
2-R-Selenophene		152, 153
R = F 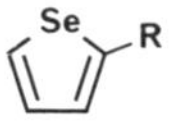	−95.4	
Cl	41.2	
Br	66.0	
I	112.6	
CH$_3$	3.8	
COOH	26.2	
CHO	−6.4	
COCH$_3$	11.7	
OCOCH$_3$	−32.4	
OCH$_3$	−91.3	
SCH$_3$	41.7	
CH(CH$_3$)OCOCH$_3$	−18.1	
CH$_2$OH	−8.6	
CON(CH$_3$)$_2$	42.3	
COOCH$_3$	24.4	
CN	104.3	
NO$_2$	5.6	
3-R-Selenophene		153
R = OCH$_3$ 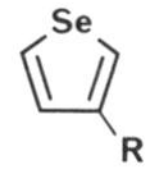	−80.7	
SCH$_3$	19.5	
Cl	13.6	
Br	38.1	
I	72.4	
CH$_3$	−15.2	
CHO	55.3	
COCH$_3$	44.8	
COOH	43.5	
CN	58.4	
NO$_2$	45.8	
Selenolo[3,2-b]selenophene[b]	−56.0	154

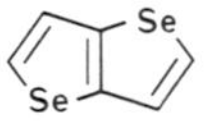

For 2-R-Selenophene: structure **(139)** with Se and R.

For 3-R-Selenophene: structure **(140)** with Se and R.

For Selenolo[3,2-b]selenophene: structure **(141)** with Se and Se.

[a]In ppm relative to degassed external selenophene. Samples were 20% solutions in acetone-d_6.
[b]20% in dimethyl-d_6 sulfoxide.

from the short relaxation times have not been fully satisfied yet. Thus ^{33}S NMR is not at present a useful tool in the study of aromatic compounds.

To some extent, it should be possible to gain understanding of the characteristics of sulfur in aromatic heterocycles by looking at the characteristics of selenium in analogous compounds. Although its NMR sensitivity is relatively low—less than half that of ^{13}C—^{77}Se is a far more tractable nucleus than ^{33}S. Its spin, $I = 1/2$, leads to reasonably narrow lines; its magnetogyric ratio is positive, which leads to useful nuclear Overhauser enhancements under proton decoupling conditions; and its measured relaxation times are helpfully long,[148-150] the shortest reported ^{77}Se T_1 being about 330 msec, for H_2Se.[151]

The largest amount of ^{77}Se NMR work on organic compounds has been done by Gronowitz and his co-workers, who have published a series of papers on substituted selenophenes[152-154] and selenides.[155] Some selenophene chemical shifts from these studies are given in Table 4.15. Even in this group of similar selenium heterocycles, the wide range and sensitivity to substituent electronic effects of ^{77}Se chemical shifts can be seen clearly. The Gronowitz group[153] attempted correlations of these substituent chemical shifts with several reactivity parameters, including the Swain and Lupton $\mathcal{F}$ and $\mathcal{R}$ constants discussed earlier in this chapter, but reasonable correlations were apparent only for the 3-substituted selenophenes, **140**.

One interesting application of ^{77}Se NMR[154] involved the determination of the structure of the high-melting point byproduct of the synthesis of selenophene from acetylene and selenium. Three structures were proposed for this byproduct—**141**, **142**, and **143**. Since only one peak was observed in the ^{77}Se

(142)   (143)
Alternative structures for **141**

spectrum and three peaks in the ^{13}C spectrum, the compound was unequivocally shown to be selenolo[3,2-b]selenophene, **141**. This structure was also most consistent with the physical properties of **141** and its sulfur analog. Recently, Gronowitz and co-workers[156] have employed ^{77}Se NMR to examine the effects of substituents at the 2-position on the selenophene **141**.

REFERENCES

1. J. B. Stothers, *Carbon-13 NMR Spectroscopy*, Academic, New York, 1972.

2. G. C. Levy and G. L. Nelson, *Carbon-13 Nuclear Magnetic Resonance for Organic Chemists*, Wiley–Interscience, New York, 1972.

3. E. Breitmaier and W. Voelter, ^{13}C *NMR Spectroscopy*, 2nd ed., Verlag Chemie, Weinheim, 1978.

4. F. W. Wehrli and T. Wirthlin, *Interpretation of Carbon-13 NMR Spectra*, Heyden, London, 1976.

5. R. J. Abraham and P. Loftus, *Proton and Carbon-13 NMR Spectroscopy: An Integrated Approach*, Heyden, London, 1978.

6. P. E. Hansen, *Org. Magn. Resonance* 12, 109 (1979).

7. W. Bremser, L. Ernst, B. Franke, R. Gerhards, and A. Hardt, *Carbon-13 NMR Spectral Data*, 2nd ed., Verlag Chemie, Weinheim, 1980.

8. *Sadtler Standard NMR Spectra*, Sadtler Research Laboratories, Philadelphia.

9. J. G. Grasselli and W. M. Ritchey (Eds.): *Atlas of Spectral Data and Physical Constants for Organic Compounds*, CRC Press, Cleveland, 1975.

10. L. F. Johnson and W. C. Jankowski, ^{13}C *NMR Spectra: A Collection of Assigned, Coded, and Indexed Spectra*, Wiley–Interscience, New York, 1972.

11. E. Breitmaier, G. Haas, and W. Voelter, *Atlas of Carbon-13 NMR Data*, Heyden, London.

12. *CNMR:* ^{13}C *NMR Spectral Search System*, NIH–EPA Chemical Information System.

13. N. K. Wilson and J. B. Stothers, *J. Magn. Resonance* 15, 31 (1974).

14. M. L. Caspar, J. B. Stothers, and N. K. Wilson, *Can. J. Chem.* 53, 1958 (1975).

15. T. D. Alger, D. M. Grant, and E. G. Paul, *J. Am. Chem. Soc.* 88, 5397 (1966).

16. J. B. Stothers, C. T. Tan, and N. K. Wilson, *Org. Magn. Resonance* 9, 408 (1977).

17. A. J. Jones and D. M. Grant, *Chem. Commun.* 1968, 1670.

18. A. J. Jones, P. J. Garratt, and K. P. C. Vollhardt, *Angew. Chem.* 85, 260 (1973).

19. N. K. Wilson, "Carbon-13 NMR Chemical Shifts and Spin-Lattice Relaxation Times of Terphenyls," paper presented at the 13th Southeastern Magnetic Resonance Conference, Durham, N.C., October, 1981. Abstract D-2.

20. J. F. M. Oth, K. Muellen, H. Koenigshofen, J. Wassen, and E. Vogel, *Helv. Chim. Acta* 57, 2387 (1974).

21. B. M. Trost and W. B. Herdle, *J. Am. Chem. Soc.* 98, 4080 (1976).

22. G. W. Buchanan and R. S. Ozubko, *Can. J. Chem.* 53, 1829 (1975).

23. R. S. Bodine, M. Hylarides, G. H. Daub, and D. L. VanderJagt, *J. Org. Chem.* 43, 4025 (1978).

24. R. S. Ozubko, G. W. Buchanan, and I. C. P. Smith, *Can. J. Chem.* 52, 2493 (1974).

25. H. Spiesecke and W. G. Schneider, *Tetrahedron Lett.* 1961, 468.

26. P. C. Lauterbur, *J. Am. Chem. Soc.* 83, 1838 (1961).

27. A. J. Jones, T. D. Alger, D. M. Grant, and W. M. Litchman, *J. Am. Chem. Soc.* 92, 2386 (1970).

28. L. Ernst, *Org. Magn. Resonance* 8, 161 (1976).

29. A. J. Jones, P. D. Gardner, D. M. Grant, W. M. Litchman, and V. Boekelheide, *J. Am. Chem. Soc.* 92, 2395 (1970).

30. D. Wendisch, W. Hartmann, and H. G. Heine, *Tetrahedron* 30, 295 (1974).

31. R. J. Pugmire, D. M. Grant, M. J. Robins, and R. K. Robins, *J. Am. Chem. Soc.* 91, 6381 (1969).

32. P. A. Claret and A. G. Osborne, *Spectrosc. Lett.* 8, 385 (1975).

33. H. Takai, A. Odani, and Y. Sasaki, *Chem. Pharm. Bull.* 26, 1672 (1978).

34. E. Breitmaier and U. Hollstein, *J. Org. Chem.* **41**, 2104 (1976).

35. U. Ewers, H. Guenther, and L. Jaenicke, *Chem. Ber.* **106**, 3951 (1973).

36. U. Ewers, H. Guenther, and L. Jaenicke, *Chem. Ber.* **107**, 876 (1974).

37. R. J. Pugmire, D. M. Grant, R. K. Robins, and G. W. Rhodes, *J. Am. Chem. Soc.* **87**, 2225 (1965).

38. R. J. Pugmire, D. M. Grant, L. B. Townsend, and R. K. Robins, *J. Am. Chem. Soc.* **95**, 2791 (1973).

39. M. C. Thorpe, W. C. Coburn, Jr., and J. A. Montgomery, *J. Magn. Resonance* **15**, 98 (1974).

40. P. Bouchet, A. Fruchier, G. Joncheray, and J. Elguero, *Org. Magn. Resonance* **9**, 716 (1977).

41. J. Elguero, A. Fruchier, and M. C. Pardo, *Can. J. Chem.* **54**, 1329 (1976).

42. C. N. Filer, F. E. Granchelli, A. H. Soloway, and J. L. Neumeyer, *J. Org. Chem.* **43**, 672 (1978).

43. A. Marker, A. J. Canty, and R. T. C. Brownlee, *Aust. J. Chem.* **31**, 1255 (1978).

44. R. J. Pugmire, M. J. Robins, D. M. Grant, and R. K. Robins, *J. Am. Chem. Soc.* **93**, 1887 (1971).

45. T. F. Page, Jr., T. Alger, and D. M. Grant, *J. Am. Chem. Soc.* **87**, 5333 (1965).

46. F. J. Weigert and J. D. Roberts, *J. Am. Chem. Soc.* **90**, 3543 (1968).

47. G. Fronza, R. Mondelli, G. Scapini, G. Ronsisvalle, and F. Vittorio, *J. Magn. Resonance* **23**, 437 (1976).

48. E. Dradi and G. Gatti, *J. Am. Chem. Soc.* **97**, 5472 (1975).

49. L. Kiezel, M. Liszka, and M. Rutkowski, *Spectrosc. Lett.* **12**, 45 (1979).

50. P. C. Lauterbur, *J. Am. Chem. Soc.* **83**, 1838 (1961).

51. R. D. Brown and M. L. Heffernan, *Aust. J. Chem.* **13**, 38 (1960).

52. G. L. Nelson and E. A. Williams, *Prog. Phys. Org. Chem.* **12**, 229 (1976).

53. V. M. Mamaev, Y. K. Grishin, and F. M. Smirnova, *Dokl. Akad. Nauk SSSR* **213**, 386 (1973).

54. V. M. Mamaev and F. M. Smirnova, *Vestn. Mosk. Univ. Khim.* **16**, 99 (1975).

55. H. R. Blattman, D. Meache, E. Heilbronner, R. J. Molyneux, and V. Boekelheide, *J. Am. Chem. Soc.* **87**, 130 (1965).

56. R. Du Vernet and V. Boekelheide, *Proc. Natl. Acad. Sci. USA* **71**, 2961 (1974).

57. C. E. Johnson, Jr. and F. A. Bovey, *J. Chem. Phys.* **29**, 1012 (1958).

58. H. Günther, H. Schmickler, H. Königshofen, K. Recker, and E. Vogt, *Angew. Chem.* **85**, 261 (1973).

59. N. K. Wilson and J. B. Stothers, *Top. Stereochem.* **8**, 1 (1974).

60. N. K. Wilson, *J. Am. Chem. Soc.* **97**, 3573 (1975).

61. D. M. Grant and B. V. Cheney, *J. Am. Chem. Soc.* **89**, 5315 (1967).

62. R. Goutarel, M. Pais, H. E. Gottlieb, and E. Wenkert, *Tetrahedron Lett.* **1978**, 1235.

63. D. F. Ewing, *Org. Magn. Resonance* **12**, 499 (1979).

64. G. L. Nelson, G. C. Levy, and J. D. Cargioli, *J. Am. Chem. Soc.* **94**, 3089 (1972).

65. G. A. Olah, P. W. Westerman, and D. A. Forsyth, *J. Am. Chem. Soc.* **97**, 3419 (1975).

66. J. A. Pople and D. L. Beveridge, *Approximate Molecular Orbital Theory*, McGraw-Hill, New York, 1970.

67. L. Ernst, *J. Magn. Resonance* **22**, 279 (1976).

68. S. Braun and J. Kinkeldei, *Tetrahedron* **33**, 1827 (1977).

69. N. K. Wilson and M. Anderson, *in* R. Haque and F. J. Biros (Eds.): *Mass Spectrometry and NMR Spectroscopy in Pesticide Chemistry*, pp. 197–218, Plenum, New York, 1974.

70. M. Witanowski, L. Stefaniak, W. Sicińska, and G. A. Webb, *J. Mol. Structure*, **64**, 15 (1980).

71. L. P. Hammett, *J. Am. Chem. Soc.* **59**, 96 (1937).

72. L. P. Hammett, *Physical Organic Chemistry*, 2nd ed., Chapter 11, McGraw-Hill, New York, 1970.

73. G. L. Anderson, R. C. Parish, and L. M. Stock, *J. Am. Chem. Soc.* **93**, 6984 (1971).

74. C. G. Swain and E. C. Lupton, Jr., *J. Am. Chem. Soc.* **90**, 4328 (1968).

75. P. R. Wells, S. Ehrenson, and R. W. Taft, *Prog. Phys. Org. Chem.* **6**, 147 (1968).

76. S. Ehrenson, R. T. C. Brownlee, and R. W. Taft, *Prog. Phys. Org. Chem.* **10**, 1 (1973).

77. J. Bromilow and R. T. C. Brownlee, *J. Org. Chem.* **44**, 1261 (1979).

78. M. J. S. Dewar, R. Golden, and J. M. Harris, *J. Am. Chem. Soc.* **93**, 4187 (1971).

79. W. F. Reynolds and G. K. Hamer, *J. Am. Chem. Soc.* **98**, 7296 (1976).

80. E M. Schulman, K. A. Christensen, D. M. Grant, and C. Walling, *J. Org. Chem.* **39**, 2686 (1974).

81. N. K. Wilson and R. D. Zehr, "Substituent Effects on the Carbon-13 NMR Spectra of 4-Substituted Para-terphenyls" paper presented at the national meeting, American Chemical Society, Washington, D.C., September 9–14, 1979. Abstract *ORGN* **217**.

82. N. K. Wilson and R. D. Zehr, *J. Org. Chem.* **47**, 1184 (1982).

83. H. E. Gottlieb, R. A. de Lima, and F. delle Monache, *J. Chem. Soc. Perkin Trans. II* **1979**, 435.

84. M. J. Shapiro, *J. Org. Chem.* **43**, 3769 (1978).

85. W. Adcock, W. Kitching, V. Alberts, G. Wickham, P. Barron, and D. Doddrell, *Org. Magn. Resonance* **10**, 47 (1977).

86. W. Kitching, M. Bullpitt, D. Gartshore, W. Adcock, T. C. Khor, D. Doddrell, and I. D. Rae, *J. Org. Chem.* **42**, 2411 (1977).

87. W. Adcock, J. Alste, S. Q. A. Rizvi, and M. Aurangzeb, *J. Am. Chem. Soc.* **98**, 1701 (1976).

88. W. Kitching, V. Alberts, W. Adcock, and D. P. Cox, *J. Org. Chem.* **43**, 4652 (1978).

89. S. N. Sawhney and D. W. Boykin, *J. Org. Chem.* **44**, 1136 (1979).

90. L. Ernst, *Z. Naturforsch.* **30b**, 794 (1975).

91. P. E. Hansen, A. Berg, H. J. Jakobsen, A. P. Manzara, and J. Michl, *Org. Magn. Resonance* **10**, 179 (1977).

92. P. E. Hansen, O. K. Poulsen, and A. Berg, *Org. Magn. Resonance* **7**, 23 (1975).

93. L. Ernst, *Z. Naturforsch.* **30b**, 800 (1975).

94. P. E. Hansen, O. K. Poulsen, and A. Berg, *Org. Magn. Resonance* **7**, 475 (1975).

95. J. Bromilow, R. T. C. Brownlee, D. J. Craik, M. Sadek, and R. W. Taft, *J. Org. Chem.* **45**, 2429 (1980).

96. J. F. Bagli and M. St.-Jacques, *Can. J. Chem.* **56**, 578 (1978).

97. T. A. Holak, S. Sadigh-Esfandiary, F. R. Carter, and D. J. Sardella, *J. Org. Chem.* **45**, 2400 (1980).

98. P. C. Lauterbur, *J. Am. Chem. Soc.* **83**, 1846 (1961).

99. P. C. Lauterbur, *J. Chem. Phys.* **38**, 1406 (1963).

100. D. M. Grant and E. G. Paul, *J. Am. Chem. Soc.* **86**, 2984 (1964).

101. J. T. Clerc, E. Pretsch, and S. Sternhell, ^{13}C-*Kernresonanzspektroskopie*, Akademische Verlagsgesellschaft, Frankfurt, 1973.

102. P. C. Lauterbur, *J. Chem. Phys.* **43**, 360 (1965).

103. N. K. Wilson and R. D. Zehr, *J. Org. Chem.* **43**, 1768 (1978).

104. I. Chu, D. C. Villeneuve, V. Secours, and A. Viau, *J. Agric. Food Chem.* **25**, 881 (1977).

105. P. R. Wells, D. P. Arnold, and D. Doddrell, *J. Chem. Soc. Perkin Trans. II* **1974**, 1745.

106. W. Kitching, M. Bullpitt, D. Doddrell, and W. Adcock, *Org. Magn. Resonance* **6**, 289 (1974).

107. M. Vajda and W. Voelter, *Z. Naturforsch.* **30**, 943 (1975).

108. D. K. Dalling, K. H. Ladner, D. M. Grant, and W. R. Woolfenden, *J. Am. Chem. Soc.* **99**, 7142 (1977).

109. J. Seita, J. Sandström, and T. Drakenberg, *Org. Magn. Resonance* **11**, 239 (1978).

110. H. Takai, A. Odani, and Y. Sasaki, *Chem. Pharm. Bull.* **26**, 1966 (1978).

111. L. Ernst, *Chem. Ber.* **108**, 2030 (1975).

112. L. Ernst, *J. Magn. Resonance* **20**, 544 (1975).

113. P. Granger and M. Maugras, *Org. Magn. Resonance* **7**, 598 (1975).

114. L. Ernst, *Z. Naturforsch.* **30b**, 788 (1975).

115. H. Fritz, T. Winkler, A. M. Braun, and C. Decker, *Helv. Chim. Acta* **61**, 661 (1978).

116. M. T. Shenbor and G. M. Smirnov, *Zh. Prikl. Spektrosk.* **28**, 885 (1978).

117. N. Platzer, J. J. Basselier, and P. Demerseman, *Bull. Soc. Chim. Fr.* **1974**, 905.

118. R. G. Parker and J. D. Roberts, *J. Org. Chem.* **35**, 996 (1970).

119. C. S. Giam and T. E. Goodwin, *J. Org. Chem.* **43**, 3780 (1978).

120. J. A. Su, E. Siew, E. V. Brown, and S. L. Smith, *Org. Magn. Resonance* **10**, 122 (1977).

121. P. A. Claret and A. G. Osborne, *Spectrosc. Lett.* **11**, 351 (1978).

122. E. Breitmaier and W. Voelter, *Tetrahedron* **30**, 3941 (1974).

123. S. Lötjönen and P. Äyräs, *Finn. Chem. Lett.* **1978**, 260.

124. N. K. Wilson, manuscript in preparation.

125. P. D. Clark, D. F. Ewing, and R. M. Scrowston, *Org. Magn. Resonance* **8**, 252 (1976).

126. D. H. Hunter and J. B. Stothers, *Can. J. Chem.* **51**, 2884 (1973).

127. *Nitrogen NMR*, M. Witanowski and G. A. Webb (Eds.), Plenum, New York, 1973.

128. G. C. Levy and R. L. Lichter, *Nitrogen-15 Nuclear Magnetic Resonance Spectroscopy*, Wiley-Interscience, New York, 1979.

129. R. O. Duthaler and J. D. Roberts, *J. Am. Chem. Soc.* **100**, 3889 (1978).

130. M. P. Sibi and R. L. Lichter, *J. Org. Chem.* **44**, 3017 (1979).

131. M. M. King, H. J. C. Yeh, and G. P. Dudek, *Org. Magn. Resonance* **8**, 208 (1976).

132. I. I. Schuster, C. Dyllick-Brenzinger, and J. D. Roberts, *J. Org. Chem.* **44**, 1765 (1979).

133. R. O. Duthaler and J. D. Roberts, *J. Am. Chem. Soc.* **100**, 4969 (1978).

134. J. P. Warren and J. D. Roberts, *J. Phys. Chem.* 78, 2507 (1974).

135. A. J. DiGioia, G. T. Furst, L. Psota, and R. L. Lichter, *J. Phys. Chem.* 82, 1644 (1978).

136. P. S. Pregosin, E. W. Randall, and A. I. White, *J. Chem. Soc. Perkin Trans. II* 1972, 1.

137. V. Markowski, G. R. Sullivan, and J. D. Roberts, *J. Am. Chem. Soc.* 99, 714 (1977).

138. I. Yavari, R. E. Botto, and J. D. Roberts, *J. Org. Chem.* 43, 2542 (1978).

139. I. I. Schuster and J. D. Roberts, *J. Org. Chem.* 44, 3864 (1979).

140. M. Filowitz, R. K. C. Ho, W. G. Klemperer, and W. Shum, *Inorg. Chem.* 18, 93 (1979).

141. B. Valentine, T. St. Amour, R. Walter, and D. Fiat, *Org. Magn. Resonance* 13, 232 (1980).

142. B. Valentine, T. St. Amour, R. Walter, and D. Fiat, *J. Magn. Resonance* 38, 413 (1980).

143. C. Delseth and J.-P. Kintzinger, *Helv. Chim. Acta* 61, 1327 (1978).

144. W. G. Klemperer, *Angew. Chem. Int. Ed. Engl.* 17, 246 (1978).

145. H. A. Christ, P. Diehl, H. R. Schneider, and H. Dahn, *Helv. Chim. Acta* 44, 865 (1961).

146. H. L. Retcofsky and R. A. Friedel, *J. Am. Chem. Soc.* 94, 6579 (1972).

147. H. L. Retcofsky and R. A. Friedel, *Appl. Spectrosc.* 24, 379 (1970).

148. W. H. Dawson and J. D. Odom, *J. Am. Chem. Soc.* 99, 8352 (1977).

149. O. D. Gansow, W. D. Vernon, and J. J. Dechter, *J. Magn. Resonance* 32, 19 (1978).

150. J. D. Odom, W. H. Dawson, and P. D. Ellis, *J. Am. Chem. Soc.* 101, 5815 (1979).

151. H. J. Jakobsen, A. J. Zozulin, P. D. Ellis, and J. D. Odom, *J. Magn. Resonance* 38, 219 (1980).

152. S. Gronowitz, I. Johnson, and A.-B. Hörnfeldt, *Chem. Scr.* 3, 94 (1973).

153. S. Gronowitz, I. Johnson, and A.-B. Hörnfeldt, *Chem. Scr.* 8, 8 (1975).

154. S. Gronowitz, T. Frejd, and A.-B. Hörnfeldt, *Chem. Scr.* 5, 236 (1974).

155. A. Fredga, S. Gronowitz, and A.-B. Hörnfeldt, *Chem. Scr.* 8, 15 (1975).

156. S. Gronowitz, A. Konar, and A.-B. Hörnfeldt, *Chem. Scr.* 19, 5 (1982).

Five

Spin–Spin Coupling Constants

As was discussed in Chapter 1, there is a source of structure in high-resolution NMR spectra that is independent of the strength of the magnetic field used in the experiment. This source of structure can be described empirically by a Hamiltonian of the form

$$\mathcal{H} = h \sum_{i < i'} J_{ii'} \mathbf{I}_i \cdot \mathbf{I}_{i'} \tag{5.1}$$

where h is Planck's constant, $\mathbf{I}_i$ and $\mathbf{I}_{i'}$ are nuclear spin operators, and $J_{ii'}$, defined by this equation, is the spin–spin coupling constant between nuclei i and i'.

More precisely, Eq. 5.1 holds for liquids in which random molecular tumbling exists; for partially oriented samples in nematic liquid crystals, the simple scalar interaction given in Eq. 5.1 must be generalized to a tensor interaction:

$$\mathcal{H} = h \sum_{i < i'} \mathbf{I}_i \cdot \mathbf{J}_{ii'} \cdot \mathbf{I}_{i'} \tag{5.2}$$

where $\mathbf{J}_{ii'}$ is now a second-rank tensor.

The origin of this interaction between pairs of nuclear spins is the coupling of electrons with each of the two nuclear spins involved. In this chapter we show how an analysis of the indirect interaction between nuclear spins through mutual couplings with the electron system gives rise to a Hamiltonian that has the form of Eq. 5.1 or 5.2, and how, with these theoretical methods, one can calculate the coupling constants $\mathbf{J}_{ii'}$, using only knowledge of the molecular structure. We then review the experimental results on spin–spin coupling constants in a variety of aromatic compounds and compare various theoretical estimates of these constants with those experimental values.

The initial discovery of spin–spin splittings was made by Gutowsky, McCall, and Slichter[1,2] and by Hahn and Maxwell;[3,4] the first basic theory of the effect was proposed by Ramsey.[5] Comprehensive reviews of the field have been made at various times by several authors, among them Barfield and Grant,[6] Murrell,[7] and Kowalewski.[8]

THEORY

The Basic Interactions

In the discussion of the basic interactions leading to nuclear spin–spin coupling, we will follow the development described by Ramsey.[5] The Hamiltonian describing the electrons in a molecule that are subject to interactions with the electric charges and magnetic moments of the nuclei in the molecule, to mutual Coulomb repulsion among themselves, and to an external magnetic field, $\mathbf{H}$, has the form

$$\mathcal{H} = \mathcal{H}_1 + \mathcal{H}_2 + \mathcal{H}_3 \tag{5.3}$$

where

$$\mathcal{H}_1 = \sum_\eta \frac{1}{2m} \left(\frac{\hbar}{i} \nabla_\eta + \frac{e}{c} \sum_i \hbar \gamma_i \mathbf{I}_i \times \frac{\mathbf{r}_{\eta i}}{\mathbf{r}_{\eta i}^3} + \frac{1}{2} \frac{e}{c} \mathbf{H} \times \mathbf{r}_\eta \right)^2$$

$$+ V + \mathcal{H}_{LL} + \mathcal{H}_{SS} + \mathcal{H}_{LS} + \mathcal{H}_{SH} \tag{5.4}$$

In this equation the electrons are indicated by η and the nuclei by i. The first term gives the kinetic energies of the electrons and their interactions, as moving charged particles, with the magnetic field of the nuclear magnetic moments and the external magnetic field, $\mathbf{H}$ (see, for example, reference 9). In Eq. 5.4, V gives the electrostatic potential energy—the sum of the interaction of the negative electrons with the positive nuclei and the mutual repulsions among themselves; $\mathcal{H}_{LS}$ is the spin orbit interaction, which arises from the motion of the electronic magnetic moment through an electrostatic field; $\mathcal{H}_{SS}$ is the direct interaction between the magnetic moments of the electrons; $\mathcal{H}_{LL}$ is the orbital interaction that gives the Hamiltonian between the electrons as moving charged particles; and $\mathcal{H}_{SH}$ represents the interaction of the electronic magnetic moments with the external field, $\mathbf{H}$. $\mathcal{H}_2$ is the classical dipole–dipole interaction between the electronic magnetic moments and the nuclear magnetic moments, and $\mathcal{H}_3$ is the Fermi contact correction to that interaction, which plays a role when the amplitude of the electron wave function is nonzero at the site of a nucleus. The forms of these interactions are[9]

$$\mathcal{H}_2 = \sum_\eta \sum_i \left[\frac{\boldsymbol{\mu}_\eta \cdot \boldsymbol{\mu}_i}{r_{\eta i}^3} - \frac{3(\boldsymbol{\mu}_\eta \cdot \mathbf{r}_{\eta i})(\boldsymbol{\mu}_i \cdot \mathbf{r}_{\eta i})}{r_{\eta i}^5} \right]$$

$$= 2\beta\hbar \sum_\eta \sum_i \gamma_i [3(\mathbf{S}_\eta \cdot \mathbf{r}_{\eta i})(\mathbf{I}_i \cdot \mathbf{r}_{\eta i}) r_{\eta i}^{-5} - (\mathbf{S}_\eta \cdot \mathbf{I}_i) r_{\eta i}^{-3}] \qquad (5.5)$$

and

$$\mathcal{H}_3 = \frac{16\pi\beta\hbar}{3} \sum_\eta \sum_i \gamma_i \delta(\mathbf{r}_{\eta i}) \mathbf{S}_\eta \cdot \mathbf{I}_i \qquad (5.6)$$

In these equations, β is the Bohr magneton, γ_i is the magnetogyric ratio of the ith nucleus, $\mathbf{S}$ and $\mathbf{I}$ are electron and nuclear spin operators, and $\delta(\mathbf{r}_{\eta i})$ is the Dirac delta function, which is zero except when $\mathbf{r}_{\eta i}$, the distance from nucleus i to electron η, is zero.

Ramsey's method was to treat all the magnetic interactions in the Hamiltonian given in Eq. 5.3 as perturbations, to use standard perturbation theory to obtain the electronic energy terms, and then to select those that involve the coordinates of two nuclear spins; he related the coefficients of those terms that were bilinear in the nuclear spins to the empirical spin-spin coupling constant given in Eq. 5.1. It is evident from an inspection of Eqs. 5.4 to 5.6 that such terms, which are bilinear in the nuclear spins, can arise from the expansion of the square of the first term in Eq. 5.4, from the direct classical dipole–dipole interaction given in Eq. 5.5, and from the Fermi contact interaction given in Eq. 5.6. The contribution from the Fermi contact interaction is typically the largest, and the computational techniques involved in evaluating that contribution to the spin-spin coupling constant are similar to those used in handling the other terms, so we will focus our attention on the details of the calculation using $\mathcal{H}_3$ alone. In general, the other terms do contribute somewhat to spin-spin coupling, and the details of the calculations using those interactions may be found in references 5 and 8.

The first-order perturbation energy resulting from the Fermi contact hyperfine interaction can be shown to vanish.[9] The second-order perturbation has the form

$$\Delta E_{(3)}^{(2)} = -\sum_i \sum_{i'} \sum_\eta \sum_{\eta'} \sum_{n(\neq 0)} \left(\frac{16\pi\beta\hbar}{3} \right)^2 \gamma_i \gamma_{i'}$$

$$\times (E_n - E_0)^{-1} (0|\delta(\mathbf{r}_{\eta i})\mathbf{S}_\eta \cdot \mathbf{I}_i|n)(n|\delta(\mathbf{r}_{\eta' i'})\mathbf{S}_{\eta'} \cdot \mathbf{I}_{i'}|0) \quad (5.7)$$

The subscript zero designates the ground state wave function and energy for the total molecular electron system; the subscript n designates excited states of the system.

If we consider only that part of $\Delta E_{(3)}^{(2)}$ depending on a particular pair of nuclei, i and i', we obtain

$$E_{(3)ii'} = -2\left(\frac{16\pi\beta\hbar}{3}\right)^2 \gamma_i\gamma_{i'} \sum_{n(\neq 0)} \sum_{\eta\eta'}$$

$$\times (E_n - E_0)^{-1}(0|\delta(\mathbf{r}_{\eta i})\mathbf{S}_\eta \cdot \mathbf{I}_i|n)(n|\delta(\mathbf{r}_{\eta'i'})\mathbf{S}_{\eta'} \cdot \mathbf{I}_{i'}|0) \quad (5.8)$$

$\mathbf{I}_i$ and $\mathbf{I}_{i'}$ can be factored out, leaving the perturbation energy as

$$E_{(3)ii'} = h\mathbf{I}_i \cdot \mathbf{J}_{ii'} \cdot \mathbf{I}_{i'} \quad (5.9)$$

where

$$\mathbf{J}_{ii'} = -\frac{2}{h}\left(\frac{16\pi\beta\hbar}{3}\right)^2 \gamma_i\gamma_{i'} \sum_{n(\neq 0)} \sum_\eta \sum_{\eta'}$$

$$\times (E_n - E_0)^{-1}(0|\delta(\mathbf{r}_{\eta i})\mathbf{S}_\eta|n)(n|\delta(\mathbf{r}_{\eta'i'})\mathbf{S}_{\eta'}|0) \quad (5.10)$$

Eq. 5.9 is precisely in the form of Eq. 5.2, and we have made one step toward our goal of finding a formula for the calculation from first principles of the spin-spin coupling constant $\mathbf{J}_{ii'}$. For molecules undergoing rapid and random rotation, it can be shown that the tensor relation in Eq. 5.9 reduces to the scalar product interaction shown in Eq. 5.1.

Having considered in some detail the perturbation correction caused by the Fermi contact interaction, we may return to a consideration of the effects of the other terms in Eqs. 5.4 to 5.6. It may be shown[8] that, in the second-order perturbation, cross terms for different perturbations cancel out for various reasons, and we are left with the following expressions for contributions to the spin-spin coupling constant from the assorted interactions:

$$J_{(1b)ii'} = -\frac{8}{3h}\beta^2 h^2 \gamma_i\gamma_{i'} \sum_n \left\langle 0\left|\sum_\eta r_{\eta i}^{-3}(\mathbf{r}_{\eta i} \times \nabla_\eta)\right|n\right\rangle$$

$$\cdot \left\langle n\left|\sum_{\eta'} r_{\eta'i'}^{-3}(\mathbf{r}_{\eta'i'} \times \nabla_{\eta i})\right|0\right\rangle \Big/ (E_n - E_0) \quad (5.11)$$

$$J_{(2)ii'} = -\frac{8}{3h}\beta^2 h^2 \gamma_i\gamma_{i'} \sum_n \left\langle 0\left|\sum_\eta 3r_{\eta i}^{-5}(\mathbf{S}_\eta \cdot \mathbf{r}_{\eta i})\mathbf{r}_{\eta i} - r_{\eta i}^{-3}\mathbf{S}_\eta\right|n\right\rangle$$

$$\cdot \left\langle n\left|\sum_{\eta'} 3r_{\eta i}^{-5}(\mathbf{S}_\eta \cdot \mathbf{r}_{\eta'i'})\mathbf{r}_{\eta'i'} - r_{\eta'i'}^{-3}\mathbf{S}_{\eta'}\right|0\right\rangle \Big/ (E_n - E_0) \quad (5.12)$$

$$J_{(3)ii'} = \frac{2}{3h}\left(\frac{16\pi\beta\hbar}{3}\right)^2 \gamma_i\gamma_{i'} \sum_n \left\langle 0\left|\sum_n \delta(\mathbf{r}_{\eta i})\mathbf{S}_\eta\right|n\right\rangle \cdot$$

$$\left\langle n \left| \sum_{\eta'} \delta(\mathbf{r}_{\eta'i'}) \mathbf{S}_{\eta'} \right| 0 \right\rangle \bigg/ (E_n - E_0) \tag{5.13}$$

$$J_{(1a)ii'} = \frac{4}{3h} \frac{e^2 h^2}{2mc^2} \gamma_i \gamma_{i'} \left\langle 0 \left| \sum_{\eta} (\mathbf{r}_{\eta i} \cdot \mathbf{r}_{\eta i'}) r_{\eta i}^{-3} r_{\eta i'}^{-3} \right| 0 \right\rangle \tag{5.14}$$

The situation becomes slightly more complicated in the case of anisotropic nematic liquid crystals, for which Eq. 5.2 holds. The contribution from $\mathcal{H}_3$ is isotropic; the generalizations of the contributions from the other mechanisms are straightforward and are discussed in reference 5. Moreover, the cross term between the spin–dipolar interaction and the Fermi contact interaction does not vanish in the anisotropic case, but gives the following contribution:

$$\mathbf{J}_{(2,3)ii'} = -\frac{64\beta^2 \hbar}{3} \gamma_i \gamma_{i'} \sum_n (E_n - E_0)^{-1} \times \left\langle 0 \left| \sum_{\eta} \delta(\mathbf{r}_{\eta i}) \mathbf{S}_\eta \right| n \right\rangle$$

$$\times \left\langle n \left| \sum_{\eta'} r_{\eta'i'}^{-5} (\mathbf{S}_{\eta'} \cdot \mathbf{r}_{\eta'i'}) \mathbf{r}_{\eta'i'} - r_{\eta'i'}^{-3} \mathbf{S}_{\eta'} \right| 0 \right\rangle \tag{5.15}$$

Use of Eqs. 5.11 to 5.15 is complicated by the fact that a knowledge of the excited states of the electron system is required. In Ramsey's original paper the "average excitation" approximation was made; in it, the differences in energy between the ground state and the excited states appearing in the denominator of the expressions were all replaced by a single average excitation energy, ΔE. In most of the more recent theories, this approximation, which is rather severe, is not made.

Molecular Orbital Theory Of The Coupling

To use Eqs. 5.11 to 5.15 for calculating spin-spin coupling constants, one needs to have expressions for the unperturbed energy states of the electronic system. Both valence bond (VB) and molecular orbital (MO) approaches to the wave function of the system have been used by different authors, with significant success.[9] The majority of the calculations that have been made in recent years, however, have been of the MO type, so we will emphasize this method of calculating spin-spin coupling constants.

McConnell[10] used MO theory with Eq. 5.14 to develop an approximate but compact and useful formula for the calculation of spin-spin coupling constants. If one begins with Eq. 5.13, approximates $(E_n - E_0)$ by ΔE for all n (the average excitation energy approximation), and uses the closure relation[9]

$$\sum_n |n\rangle\langle n| = 1 \tag{5.16}$$

one obtains

$$J_{ii'} \approx -\frac{2}{3h}\left(\frac{16\pi\beta\hbar}{3}\right)^2 \gamma_i\gamma_{i'}(\Delta E)^{-1} \sum_\eta\sum_{\eta'}\sum_n \times (0|\delta(\mathbf{r}_{\eta i})\mathbf{S}_\eta|n) \cdot (n|\delta(\mathbf{r}_{\eta' i'})\mathbf{S}_{\eta'}|0)$$

$$= -\frac{2}{3h}\left(\frac{16\pi\beta\hbar}{3}\right)^2 \gamma_i\gamma_{i'}(\Delta E)^{-1} \sum_\eta\sum_{\eta'} (0|\delta(\mathbf{r}_{\eta i})\delta(\mathbf{r}_{\eta' i'})\mathbf{S}_\eta \cdot \mathbf{S}_{\eta'}|0) \tag{5.17}$$

To construct the ground state wave function, McConnell used single-electron MO's expressed as a linear combination of atomic orbitals (LCAO) in the standard fashion,

$$\psi_J = \sum_j C_{Jj}\phi_j, \tag{5.18}$$

and a closed-shell singlet ground state, with each molecular orbital doubly occupied by electrons of opposite spin. Inserting this ground state wave function into Eq. 5.17 and carrying out the indicated integrals, he obtained

$$J_{ii'} = \tfrac{16}{9}\beta^2 h\gamma_i\gamma_{i'}(\Delta E)^{-1}\phi(0)^4 P_{ii'}^2 \tag{5.19}$$

for the spin–spin coupling constant between nuclei i and i'; $P_{ii'}$ is the usual bond order

$$P_{ii'} = 2\sum_J C_{Ji}C_{Ji'} = 2\sum_K C_{Ki}C_{Ki'} \tag{5.20}$$

and $\phi(0)$ is the amplitude of an s atomic orbital at the origin. It is the presence of the Dirac delta function in Eq. 5.6 that results in the atomic orbital's being evaluated at the site of the nucleus. Since only s state atomic orbitals have non-zero amplitude at the origin, these are the ones that appear in the final expression. Eq. 5.19 had significant semiquantitative success in accounting for general trends in experimental results concerning spin–spin coupling constants, but was limited in accuracy and, since it is clearly always positive from its mathematical form, was unable to account for the observation of negative coupling constants.

A useful generalization of this method was developed by Pople and Santry.[11] They did not use the average energy approximation and so were unable to use the closure relation to sidestep the necessity of having knowledge of the excited states of the system. To approximate using all the excited states of the system, they considered only the abbreviated set of excited states consisting of states in which a single electron is promoted from its ground state MO to an excited state MO, and where only a finite number of excited single state molecular orbitals are considered. Moreover, they considered only electron integrals involving a single atomic center. With these approximations, the expression for the coupling constant becomes

$$J_{(3)ii'} = -\tfrac{64}{9}\beta^2 h\gamma_i\gamma_{i'}\phi_i^2(0)\phi_{i'}^2(0)\Pi_{ii'} \tag{5.21}$$

where

$$\Pi_{ii'} = -4 \sum_{J}^{occ} \sum_{K}^{unocc} \frac{C_{Ji} C_{Ki} C_{Ji'} C_{Ki'}}{E_K - E_J} \qquad (5.22)$$

defines the atom-atom polarizability. E_J is the energy of an occupied MO, and E_K that of an unoccupied MO. It can be shown that if the average energy approximation is invoked, Eq. 5.21 reduces to the McConnell expression given in Eq. 5.19. The Pople-Santry equation has had considerable success and is still frequently used.

The question now arises as to which forms of MO theory are to be used in calculating the MO coefficients, C_{Ji}, and energies, E_J. Several methods of varying computational difficulty and accuracy have been used. The simplest method, and the least accurate, is a generalization of the Hückel theory of π-electron systems, in which the electrons are considered to be independent of one another, but which is extended to include all valence electrons rather than π electrons only; Hoffman's extended Hückel theory (EHT)[12] and a technique developed by Pople and Santry[13] are examples of this approach. More accurate are the semiempirical self-consistent field (SCF) methods that include all valence electrons—the Hartree-Fock equations, which incorporate the Coulomb repulsion between the electrons and which are to be solved iteratively. The basic method as applied to molecular orbitals was developed by Roothaan.[14] Two forms of this theory, differing in the treatment of certain molecular integrals involved in the calculation, are the complete neglect of differential overlap (CNDO) and the intermediate neglect of differential overlap (INDO) methods of Pople and his colleagues.[15-19] Of these two, the INDO method is the less approximate. Finally, there are nonempirical (*ab initio*) SCF methods in which the integral approximations involved in the semi-empirical SCF methods described previously are not made.[14] Calculational procedures also differ in whether Slater-type orbitals (STO) or Gaussian type orbitals (GTO) are used for the atomic orbitals.

Further improvements in the accuracy of the unperturbed wave functions entering into Eq. 5.21 require configuration interaction (CI) methods to account adequately for electron correlation.[20-22] In these calculations, linear combinations of total electron excited state wave functions are used in a variational procedure. See reference 8 for details.

The Finite Perturbation Method

In all the methods based on Eq. 5.21, which are collectively called sum-over-states (SOS) methods, the wave functions are calculated in the absence of any interaction involving either nuclear or electron spin. There is an alternative way to proceed in which the Hartree-Fock SCF iterative procedure is used but in which the terms in the Hamiltonian involving nuclear and electron spin coordinates are included in the initial calculational procedure. These methods are called coupled Hartree-Fock methods and include two techniques for the calculation of spin-spin coupling constants—the finite perturbation (FP) method of Pople, McIver, and Ostlund,[23-26] and the self-consistent perturbation theory (SCPT) of Blizzard and Santry.[27,28] The first of these is perhaps currently the most widely used technique for the calculation of spin-spin coupling constants.

In the coupled Hartree-Fock approach to calculating the Fermi contact contribution to $J_{ii'}$, one writes the Hamiltonian

$$\mathcal{H} = \mathcal{H}_0 + \mu_i \mathcal{H}_i + \mu_{i'} \mathcal{H}_{i'} \tag{5.23}$$

where

$$\mu_i = \gamma \hbar I_i \tag{5.24}$$

and

$$\mathcal{H}_i = \frac{16\pi\beta}{3} \sum_\eta \delta(\mathbf{r}_{\eta i}) S_{z\eta} \tag{5.25}$$

This formulation is equivalent to that given earlier.

For convenience, one then defines a reduced coupling constant, $K_{ii'}$, in terms of $J_{ii'}$ as

$$K_{ii'} = 4\pi^2 J_{ii'}/(h\gamma_i\gamma_{i'}) \tag{5.26}$$

so that, in direct analogy with the procedure given before, we have

$$K_{ii'} = \left[\frac{\partial^2 E(\mu_i, \mu_{i'})}{\partial\mu_i\partial\mu_{i'}}\right]_{\mu_i = \mu_{i'} = 0} \tag{5.27}$$

One evaluates the derivative appearing in Eq. 5.27 by using the Hellman-Feynman theorem:[23]

$$K_{(3)ii'} = \left[\frac{\partial}{\partial\mu_{i'}} \langle\Psi(\mu_{i'})|\mathcal{H}_i|\Psi(\mu_{i'})\rangle\right]_{\mu_{i'}=0} \tag{5.28}$$

where $\Psi(\mu_{i'})$ is the wave function of the system calculated *in the presence of* $\mu_{i'}$.

It can now be shown that the expectation value appearing in Eq. 5.28 may be written in terms of the spin density as follows:

$$\langle\Psi(\mu_{i'})|\mathcal{H}_i|\Psi(\mu_{i'})\rangle = \frac{8\pi\beta}{3} \sum_{jk} \rho_{jk}(\mu_{i'})\langle j|\delta(\mathbf{r}_i)|k\rangle \tag{5.29}$$

The spin density, ρ, is a measure of the difference of wave function density for electrons associated with spin up and spin down; the presence of the nuclear spin magnetic moment appearing in the Hamiltonian in Eq. 5.23 means that one electron orientation is energetically preferable to the other spin orientation. Combining Eqs. 5.28 and 5.29, we obtain

$$K_{(3)ii'} = \frac{8\pi\beta}{3} \sum_{jk} \langle j|\delta(\mathbf{r}_i)|k\rangle \left[\frac{\partial}{\partial\mu_{i'}} \rho_{jk}(\mu_{i'})\right]_{\mu_{i'}=0} \tag{5.30}$$

where the sum over j and k is over pairs of atomic orbitals.

In the FP method, integral approximations equivalent to either the CNDO or INDO method are introduced and the derivative appearing in Eq. 5.30 is evaluated by finite difference methods. The final expression for the reduced coupling constant is

$$K_{(3)ii'} = \left(\frac{8\pi\beta}{3}\right)^2 \phi_i^2(0)\, \phi_{i'}^2(0)\, \frac{\rho_{ii'}(h_{i'})}{h_{i'}} \tag{5.31}$$

$\rho_{ii'}(h_{i'})$ is to be calculated by SCF methods; $h_{i'}$ is a parameter entering in the finite difference treatment of the derivative appearing in Eq. 5.30. The magnitude of $h_{i'}$ may be chosen so as to maximize the accuracy of the method[24] and is of the order of 10^{-3}. A program (QCPE 281) to perform FP calculations of coupling constants is available from the Quantum Chemistry Program Exchange, Chemistry Department, Indiana University.

In the SCPT method of Blizzard and Santry,[27,28] the derivative appearing in Eq. 5.28 is calculated from first-order corrections to the MO's rather than by use of a finite difference technique. The final result in this method is

$$K_{(3)ii'} = 2P_{ii'}^{(1)} \left(\frac{8\pi\beta}{3}\right)^2 \phi_i^2(0)\, \phi_{i'}^2(0) \tag{5.32}$$

where

$$P_{jk}^{(1)} = \frac{1}{2}\left(\frac{\partial \rho_{jk}}{\partial \mu_{i'}}\right)_{\mu_{i'}=0} \tag{5.33}$$

Detailed comparisons between Eqs. 5.21, 5.31, and 5.32 have been carried out by a number of authors.[25,29,30] The methods are fundamentally similar and do not differ greatly in validity. Some further refinements in accuracy have depended on CI methods involving more than one Slater determinant.[8] A number of calculations of spin–spin coupling constants have been made using the double perturbation technique, in which the nuclear-electron spin interaction and, independently, the effects of electron correlation are considered as separate perturbations.[31-34]

The π-Electron Contribution to Spin–Spin Coupling

Any consideration of coupling constants in aromatic compounds must deal with the difference between sigma and π-electron contributions to the effect; the π-electron contributions are particularly important for long-range coupling. There is an apparent paradox involved: π electrons have zero density in the plane of an aromatic ring and so would not seem to be able to contribute to spin–spin coupling through the Fermi contact interaction. McConnell,[35] however, suggested that electron exchange involving sigma and π electrons can allow π electrons to contribute to spin–spin coupling in much the same way that they contribute to electron–proton hyperfine coupling in π radicals. To develop formulas that relate the observed constants, several authors have made use of the fact that this interaction with the sigma electrons, through which the π electrons make themselves felt, is the same for nuclear spin-spin coupling and ESR hyperfine coupling. One of these formulas, based on MO methods, is

$$J'_{HH} = \frac{Q_{CH} Q_{C'H'} \pi_{CC}}{4} \tag{5.34}$$

where Q_{CH} is the experimental hyperfine coupling constant and π_{CC} is the usual atom-atom polarizability.[36,37] Theories that calculate the π-electron contribution to spin-spin coupling without using the empirical ESR hyperfine coupling constants have been developed.[8,38,39]

One interesting observation is that the FP-INDO calculations and FP-CNDO calculations differ in the exclusion in CNDO of the exchange integrals that incorporate the mechanism by which the π electrons contribute to the spin-spin coupling, so that looking at the difference in the coupling constant calculated by the two methods should give a measure of the π-electron contribution.[25] A further important observation, attributed to Hoffman,[40] suggests that if the π-electron system transmits coupling involving a proton, and if a methyl group replaces the proton, the new coupling constant should be of comparable magnitude but opposite sign.

EXPERIMENTAL RESULTS

In the previous section we summarized theoretical methods that have been developed to calculate spin-spin coupling constants. The most widely used formulas are Eqs. 5.21 and 5.31, particularly the latter. In the rest of this chapter, we will give experimental results on coupling constants in aromatic compounds, and we will compare experimental values and theoretical values of those coupling constants calculated by the methods described in this chapter.

Tables 5.1 to 5.8 present experimental results obtained by a number of au-

Table 5.1. H–H Spin-Spin Coupling Constants (in Hertz) of Benzenoid Hydrocarbons (J_{th} are FP-INDO values given in reference 81; J_{ex} are experimental values.)

Molecule	Bond	J_{th}	J_{ex}
Benzene	H_1-H_2	+8.31	$7.48^a, 7.47^b, +7.54^c, 7.52^d, 7.56^c$
	H_1-H_3	+2.15	$1.32^a, 1.31^b, +1.37^c, 1.31^d, 1.38^c$
	H_1-H_4	+1.17	$0.596^a, 0.66^b, +0.69^c, 0.65^d, 0.69^c$
	H_1-H_5	+2.15	$1.32^a, 1.31^b, +1.37^c, 1.31^d, 1.38^c$
	H_1-H_6	+8.31	$7.48^a, 7.47^b, +7.54^c, 7.52^d, 7.56^c$

(1)

Table 5.1. (Continued)

Molecule	Bond	J_{th}	J_{ex}
Naphthalene	H_1-H_2	+8.18	$8.2^f, 8.6^g, 8.1^h, 8.2^i$
	H_1-H_3	+1.98	$1.2^f, 1.4^g, 1.1^h, +1.25^j$
	H_1-H_4	+1.80	0.0^f
	H_1-H_5	+1.81	$+0.83^j$
	H_1-H_6	−0.436	-0.16^j
	H_1-H_7	+0.541	$+0.21^j$
	H_2-H_1	+8.19	$8.2^f, 8.6^g, 8.1^h, 8.2^i$
	H_2-H_3	+8.82	$6.7^f, 6.4^h, 6.9^i$
	H_2-H_4	+1.97	$1.2^f, 1.4^g, 1.1^h, +1.25^j$
	H_2-H_6	+0.411	0.28^j

(2)

Molecule	Bond	J_{th}	J_{ex}
Anthracene	H_1-H_2	+8.36	8.3^h
	H_1-H_3	+1.84	1.2^h
	H_2-H_1	+8.36	8.3^h
	H_2-H_3	+8.94	6.5^h

(3)

Molecule	Bond	J_{th}	J_{ex}
Phenanthrene	H_1-H_2	+8.14	$8.64^k, 8.23^k, 8.4^l$
	H_1-H_3	+2.07	$1.07^k, 0.95^k, 1.6^l$
	H_1-H_4	+1.68	$0.60^k, 0.65^k, 0.5^l$
	H_2-H_1	+8.14	$8.64^k, 8.23^k, 8.4^l$
	H_2-H_3	+8.70	$7.25^k, 6.83^k, 7.3^l$
	H_2-H_4	+2.18	$1.38^k, 1.52^k, 1.6^l$
	H_3-H_1	+2.06	$1.07^k, 0.95^k, 1.6^l$
	H_3-H_2	+8.71	$7.25^k, 6.83^k, 7.3^l$
	H_3-H_4	+8.18	$8.17^k, 7.84^k, 8.4^l$
	H_4-H_1	+1.68	$0.60^k, 0.65^k, 0.5^l$
	H_4-H_2	+2.18	$1.38^k, 1.52^k, 1.6^l$
	H_4-H_3	+8.16	$8.17^k, 7.84^k, 8.4^l$

(4)

Molecule	Bond	J_{th}	J_{ex}
Pyrene	H_1-H_2	+8.26	$7.6^h, 7.75^m, 7.75^m$
	H_1-H_3	+2.10	$1.09^m, 1.00^m$
	H_2-H_1	+8.26	$7.6^h, 7.75^m, 7.75^m$
	H_2-H_3	+8.19	$7.6^h, 7.75^m, 7.75^m$
	H_4-H_5	+8.57	8.97^m

(5)

[a]Reference 41. [e]Reference 45. [h]Reference 48. [k]Reference 51.

[b]Reference 42. [f]Reference 46. [i]Reference 49. [l]Reference 52.

[c]Reference 43. [g]Reference 47. [j]Reference 50. [m]Reference 53.

[d]Reference 44.

Table 5.2. C–H Spin–Spin Coupling Constants (in Hertz) in Benzenoid Hydrocarbons (J_{th} are FP–INDO values given in reference 81; J_{ex} are experimental values.)

Molecule	Bond	J_{th}	J_{ex}
Benzene, 1	C_1-H_1	+140	+159[a], 158[b], 159[c]
	C_2-H_1	−5.01	+1.11[a], +1.0[b]
	C_3-H_1	+9.44	+7.58[a], +7.4[b]
	C_4-H_1	−2.27	−1.20[a], −1.1[b]
	C_5-H_1	+9.44	+7.58[a], +7.4[b]
	C_6-H_1	−5.01	+1.11[a], +1.0[b]
Naphthalene, 2	C_1-H_1	+146	1.58[d], 158[e]
	C_2-H_1	−5.51	1.7[d]
	C_3-H_1	+10.1	8.6[d]
	C_4-H_1	−3.42	−1.5[d]
	C_8-H_1	+6.47	4.6[d]
	C_1-H_2	−6.61	0.2[d]
	C_2-H_2	+139	158[d], 158[e]
	C_3-II_2	−5.52	0.3[d]
	C_4-H_2	+9.44	7.6[d]
Biphenyl	C_1-H_1	+140	162[e]
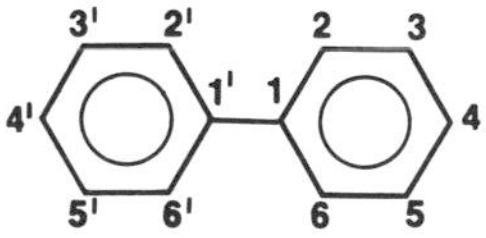	C_2-H_2	+139	162[e]
	C_3-H_3	+151	162[e]
Phenanthrene, 4	C_1-H_1	+145	158[e]
	C_2-H_2	+139	158[e]
	C_3-H_3	+139	158[e]
	C_4-H_4	+148	158[e]
	$C_{10}-H_{10}$	+145	158[e]
Pyrene, 5	C_1-H_1	+146	154[e], 159[f], 159[f]
	C_2-H_2	+136	154[e], 159[f], 159[f]
	C_4-H_4	+146	154[e], 158[f]

Biphenyl structure labeled with positions 3', 2', 2, 3; 4', 1', 1, 4; 5', 6', 6, 5. (6)

[a]Reference 54. [d]Reference 57.
[b]Reference 55. [e]Reference 58.
[c]Reference 56. [f]Reference 53.

Table 5.3. C–C Spin–Spin Coupling Constants (in Hertz) of Benzenoid Hydrocarbons (J_{th} are FP-INDO values given in reference 81; J_{ex} are experimental values.)

Molecule	Bond	J_{th}	J_{ex}
Benzene, 1	C_1-C_2	+75.9	57.0[a]
	C_1-C_6	+75.9	57.0[a]
Naphthalene, 2	C_2-C_1	+74.8	60.3[b]
	C_2-C_4	−12.4	2.43[b]
	C_2-C_5	−4.80	1.47[b]
	C_2-C_8	+8.58	5.45[b]
	C_2-C_9	−8.99	1.69[b]
	C_2-C_{10}	+10.1	7.97[b]
Pyrene, 5	C_1-C_2	+74.9	57[b]
	C_1-C_4	−7.34	1.55[b]
	C_1-C_5	+7.20	2.27[b]
	C_1-C_7	−4.80	0.76[b]
	C_1-C_9	+11.4	5.82[b]
	C_1-C_{10}	−12.4	1.55[b]
	C_1-C_{10a}	+70.3	58.9[b]
	C_1-C_{3a}	+13.1	7.71[b]
	C_1-C_{5a}	−4.65	1.35[b]
	C_1-C_{8a}	−5.42	0.31[b]
	C_1-C_{10c}	+6.97	3.05[b]
	C_1-C_{10b}	−9.65	0.2[b]

[a] Reference 59.
[b] Reference 60.

Table 5.4. H–H Spin–Spin Coupling Constants (in Hertz) for Nitrogen Heterocyclics (J_{th} are FP-INDO values given in reference 82; J_{ex} are experimental values.)

Molecule	Bond	J_{th}	J_{ex}
Pyridine	H_2-H_3	+5.27	4.87[a], 4.88[b], 4.97[b], 4.86[c]
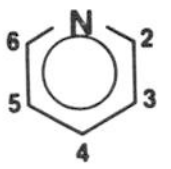	H_2-H_4	+2.36	1.85[a], 1.83[b], 1.81[b], 1.85[c]
	H_2-H_5	+2.22	1.01[a], 0.97[b], 0.90[b], 0.98[c]
	H_2-H_6	+0.0318	−0.15[a], −0.12[b], −0.16[b], −0.13[c]
(7)	H_3-H_4	+9.21	7.65[a], 7.62[b], 7.83[b], 7.66[c]
	H_3-H_5	+1.60	1.35[a], 1.34[b], 1.38[b], 1.36[c]
	H_3-H_6	+2.23	1.01[a], 0.97[b], 0.90[b], 0.98[c]
	H_4-H_5	+9.21	7.65[a], 7.62[b], 7.83[b], 7.66[c]
	H_4-H_6	+2.37	1.85[a], 1.83[b], 1.81[b], 1.85[c]

Table 5.4. (Continued)

Molecule	Bond	J_{th}	J_{ex}
Pyridazine	H_3-H_4	+6.16	+5.05[d], 5.2[e], 4.9[f]
	H_3-H_5	+1.55	+1.88[d], 1.9[e], 2.0[f]
	H_3-H_6	+3.41	+1.39[d], 1.4[e], 3.5[f]
	H_4-H_5	+10.2	+8.22[d], 8.6[e], 8.4[f]
	H_4-H_6	+1.49	+1.88[d], 1.9[e], 2.0[f]
(8)			
Pyrimidine	H_2-H_4	+0.287	~0[g]
	H_2-H_5	+3.08	1.5[g]
	H_2-H_6	+0.304	~0[g]
	H_4-H_5	+5.79	5.0[g]
	H_4-H_6	+2.87	2.5[g]
(9)	H_5-H_6	+6.06	5.0[g]
Pyrazine	H_2-H_3	+3.33	1.8[f]
	H_2-H_5	+3.14	1.8[f]
	H_2-H_6	-0.549	~0.5[f]
(10)			
Quinoline	H_2-H_3	+5.95	4.25[h], 4.25[h], 4.19[h]
	H_2-H_4	+1.72	1.79[h], 1.77[h], 1.75[h]
	H_3-H_4	+9.45	8.32[h], 8.34[h], 8.28[h]
	H_4-H_5	-0.797	0.4[i]
	H_4-H_8	+2.30	0.9[h], 0.8[h], 0.9[h]
(11)	H_5-H_6	+8.15	8.15[h], 8.07[h], 8.22[h], 8.30[j], 8.15[j]
	H_5-H_7	+2.17	1.60[h], 1.49[h], 1.46[h], 1.45[j], 1.45[j]
	H_5-H_8	+1.80	0.33[h], 0.48[h], 0.55[h], 0.65[j], 0.70[j]
	H_6-H_7	+8.62	6.81[h], 6.82[h], 6.92[h], 6.97[j], 6.92[j]
	H_6-H_8	+1.67	1.12[h], 1.10[h], 1.16[h], 1.6[j], 1.20[j]
	H_7-H_8	+8.99	8.27[h], 8.41[h], 8.62[h], 8.4[j], 8.55[j]
Quinoxaline	H_2-H_7	-0.339	>0[k], >0[k]
	H_5-H_6	+8.84	8.6[j], 8.40[k], 8.40[k], 8.45[l]
	H_5-H_7	+2.03	1.4[j], 1.46[k], 1.55[k]
	H_5-H_8	+1.61	0.8[j], 0.60[k], 0.58[k]
	H_6-H_7	+8.33	7.1[j], 6.94[k], 6.89[k], 6.94[l]
(12)	H_6-H_8	+2.03	1.4[j], 1.46[k], 1.55[k]
Phthalazine	H_1-H_5	+2.02	~0.4[k]
	H_5-H_6	+8.13	8.17[k]
	H_5-H_7	+1.91	1.24[k]
	H_5-H_8	+1.84	0.57[k]
	H_6-H_7	+8.85	6.76[k]
(13)	H_6-H_8	+1.91	1.24[k]

Table 5.4. (Continued)

Molecule	Bond	J_{th}	J_{ex}
Isoquinoline	H_1-H_3	-0.428	$\sim0^m, \sim0^m$
(14)	H_1-H_5	+2.06	$<0.5^m, <0.5^m$
	H_3-H_4	+5.35	$5.8^m, 6.0^m$
	H_4-H_8	+2.10	$\sim0.8^m, \sim0.8^m$
	H_5-H_6	+8.31	$8.62^m, 8.68^m$
	H_5-H_7	+1.81	$0.88^m, 1.07^m$
	H_5-H_8	+1.98	$0.82^m, 0.90^m$
	H_6-H_7	+8.94	$7.02^m, 6.99^m$
	H_6-H_8	+1.79	$1.09^m, 1.29^m$
	H_7-H_8	+8.30	$8.39^m, 8.21^m$
Cinnoline	H_3-H_4	+6.88	$5.75^k, 5.80^k$
(15)	H_4-H_8	+2.79	0.83^k
	H_5-H_6	+8.25	7.87^k
	H_5-H_7	+1.98	1.57^k
	H_5-H_8	+1.97	0.85^k
	H_6-H_7	+8.74	6.94^k
	H_6-H_8	+1.50	1.34^k
	H_7-H_8	+9.04	8.64^k
Quinazoline	H_2-H_4	-0.133	$\sim0^k, \sim0^k$
(16)	H_4-H_8	+2.29	$\sim0.5^k, >0^k$
	H_5-H_6	+8.11	$8.21^j, 8.23^k, 7.99^k$
	H_5-H_7	+2.18	$1.45^j, 1.37^k, 1.26^k$
	H_5-H_8	+1.80	$0.70^j, 0.63^k, 0.89^k$
	H_6-H_7	+8.63	$6.71^j, 7.03^k, 6.99^k$
	H_6-H_8	+1.66	$1.43^j, 0.89^k, 1.26^k$
	H_7-H_8	+8.95	$8.40^j, 8.59^k, 8.58^k$
Acridine	H_1-H_2	+9.37	9.0^n
(17)	H_1-H_3	+1.37	1.2^n
	H_1-H_4	+2.24	0.6^n
	H_1-H_{10}	+3.35	0.9^n
	H_2-H_3	+8.95	6.6^n
	H_2-H_4	+1.79	1.4^n
	H_3-H_4	+8.53	8.2^n
	H_4-H_{10}	-1.76	0.4^n
Phenazine	H_1-H_2	+10.4	8.80^l
(18)	H_2-H_3	+9.66	6.63^l

Table 5.4. (Continued)

Molecule	Bond	J_{th}	J_{ex}
Benzo(g) quinoxaline	H_7–H_8	+8.41	8.63[l]

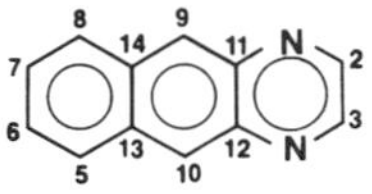

(19)

Molecule	Bond	J_{th}	J_{ex}
Benzo(b) phenazine	H_1–H_2	+12.4	8.94[l]
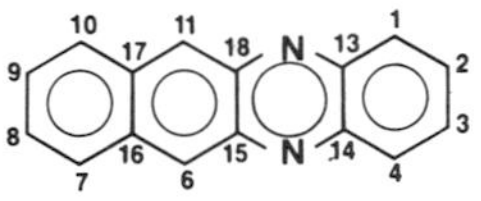	H_2–H_3	+11.5	6.39[l]
	H_8–H_9	+9.94	6.45[l]
	H_9–H_{10}	+9.55	8.79[l]

(20)

[a]Reference 61. [e]Reference 65. [i]Reference 69. [l]Reference 72.
[b]Reference 62. [f]Reference 66. [j]Reference 70. [m]Reference 73.
[c]Reference 63. [g]Reference 67. [k]Reference 71. [n]Reference 74.
[d]Reference 64. [h]Reference 68.

Table 5.5. C–H Spin–Spin Coupling Constants (in Hertz) for Nitrogen Heterocyclics

Molecule	Bond	J_{th}	J_{ex}
Pyridine, 7	C_2–H_2	+173	177[a], 178[a], 176[a], 175[b], +178[c], 180[d], 179[e], 170[f]
	C_2–H_3	-6.24	4.6[a], 3.3[b], +3.12[c]
	C_2–H_4	+10.1	6.7[a], 6.4[b], +6.85[c]
	C_2–H_5	-3.92	<0.8[a], ±1.6[b], -0.92[c]
	C_2–H_6	+13.0	11.3[a], 10.9[b], +11.2[c]
	C_3–H_2	-0.666	8.3[a], 8.7[b], +8.47[c]
	C_3–H_3	+141	162[a], 163[a], 162[a], 163[b], +163[c], 162[d], 163[e], 163[f]
	C_3–H_4	-6.17	<0.5[a], 1.0[b], +0.84[c]
	C_3–H_5	+8.87	6.7[a], 6.4[b], +6.56[c]
	C_3–H_6	-2.84	<2.0[a], ±1.6[b], -1.65[c]
	C_4–H_2	+8.55	6.2[a], 6.4[b], +6.34[c]
	C_4–H_3	-6.18	<1.0[a], 0.0[b], +0.70[c]
	C_4–H_4	+142	163[a], 161[a], 163[a], 169[b], +162[c], 160[d], 152[e], 152[f]
	C_4–H_5	-6.18	<1.0[a], 0.0[b], +0.70[c]
	C_4–H_6	+8.55	6.2[a], 6.4[b], +6.34[c]

Table 5.5. (Continued)

Molecule	Bond	J_{th}	J_{ex}
Pyridazine, **8**	C_3-H_3	+173	183[b], 186[d], 182[e], 182[g]
	C_3-H_4	−8.47	6.5[b]
	C_3-H_5	+9.72	2.0[b]
	C_3-H_6	−5.12	−1.4[b]
	C_4-H_3	−3.38	6.7[b]
	C_4-H_4	+143	170[b], 174[d], 169[e], 169[g]
	C_4-H_5	−7.90	0.0[b]
	C_4-H_6	+8.79	5.2[b]
Pyrimidine, **9**	C_2-H_2	+209	203[b], 211[d], 206[e], 206[h]
	C_2-H_4	+13.2	10.3[b]
	C_2-H_5	−6.12	0.0[b]
	C_2-H_6	+14.7	10.3[b]
	C_4-H_2	+12.3	9.1[b]
	C_4-H_4	+174	183[b], 182[d], 182[e], 182[h]
	C_4-H_5	−8.04	1.9[b]
	C_4-H_6	+10.1	5.3[b]
	C_5-H_2	−4.44	1.9[b]
	C_5-H_4	−2.71	9.5[b]
	C_5-H_5	+142	166[b], 171[d], 168[e], 168[h]
	C_5-H_6	−2.72	9.5[b]
Pyrazine, **10**	C_2-H_2	+173	183[b], 184[d], 183[e], 183[g]
	C_2-H_3	−1.38	10.4[b]
	C_2-H_5	−4.77	−1.5[b]
	C_2-H_6	+12.1	9.8[b]
s-Triazine	C_2-H_2	+209	208[b], 206[d], 208[e]
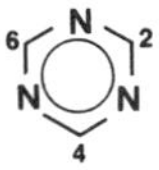	C_2-H_4	+12.0	7.95[b]
	C_2-H_6	+11.9	7.95[b]
(21)			
Quinoxaline, **12**	C_2-H_2	+174	182[i]
	C_2-H_3	−4.79	11.5[i]
	C_5-H_5	+138	162[i]
	C_6-H_6	+142	162[i]
Phthalazine, **13**	C_1-H_1	+177	181[i]
	C_5-H_5	+147	164[i]

Source. Reference 82.

[a]Reference 75.   [d]Reference 77.   [g]Reference 66.
[b]Reference 76.   [e]Reference 78.   [h]Reference 67.
[c]Reference 61.   [f]Reference 79.   [i]Reference 80.

Table 5.6. C–C Spin–Spin Coupling Constants (in Hertz) for Pyridine, 7 (J_{th} are FP-INDO values from reference 82; J_{ex} are experimental values.)

Bond	J_{th}	J_{ex}
$C_2–C_3$	+80.6	53.8[a]
$C_2–C_5$	+15.9	13.9[a]
$C_3–C_4$	+75.4	53.5[a]

Source. Reference 82.
[a]Reference 101.

Table 5.7. N–H Spin–Spin Coupling Constants (in Hertz) for Nitrogen Heterocyclics (J_{th} are FP-INDO values given in reference 82; J_{ex} are experimental values.)

Molecule	Bond	J_{th}	J_{ex}
Pyridine, 7	$N_1–H_2$	+10.8	−10.8[a], −10.1[a]
	$N_1–H_3$	+0.826	−1.53[a], −1.56[a]
	$N_1–H_4$	−0.789	±0.21[a], ±0.18[a]
Pyridazine, 8	$N_2–H_3$	+7.57	−12.0[b]
	$N_2–H_4$	+0.862	−1.12[b]
	$N_2–H_5$	−0.650	−0.36[b]
	$N_2–H_6$	+1.58	−3.70[b]
Quinoline, 11	$N_1–H_2$	+8.08	−11.0[c]
	$N_1–H_3$	+1.72	−1.3[c]

Source. Reference 82.   [b]Reference 64.
[a]Reference 62.   [c]Reference 100.

Table 5.8. N–C Spin–Spin Coupling Constants (in Hertz) for Pyridine, 7 (J_{th} are FP-INDO values given in reference 82; J_{ex} are experimental values.)

Bond	J_{th}	J_{ex}
$N_1–C_2$	+2.21	±0.45[a], ±0.7[a]
$N_1–C_3$	−5.17	±2.4[a], ±2.6[a]
$N_1–C_4$	+5.87	±3.6[a], ±3.8[a]

Source. Reference 82.
[a]Reference 62.

Table 5.9. ^{13}C–^{13}C Spin–Spin Coupling Constants in Some Monosubstituted Benzenes[a]

X	1, 2	1, 3	1, 4	2, 3	2, 4	2, 5
F	70.79		10.47	56.64	2.89	6.68
Cl	65.16[d]		10.66[c]	55.77[c]	2.81	7.80
Br	63.65		10.66	54.86[c]	2.75[e]	8.13
I	60.93	2.53	10.62	54.44	2.63	8.72
OH	65.6		9.6	57.6		7.1
OCH$_3$	67.04		9.18	57.76	2.81	7.13
NO$_2$	67.38[e]		9.70[e]	56.11	2.63[e]	7.39
NH$_2$	61.1		9.2[e]	58.6		7.9
NHCH$_3$	61.7		8.6[e]	59.1		7.9
N(CH$_3$)$_2$	62.84		8.18[e]	58.99	2.74	7.98
NHNH$_2$	62.7			58.6		
NCO	65.9		11.0	57.0		7.9
NCS	66.0[e]			57.0[d]		
COCH$_3$	57.8					
CO$_2$H				57.4[c]		
CO$_2$CH$_3$	58.3[e]		9.4	56.3[c]		
CHO	57.9		8.9			8.8
CFO	59.2			56.4[c]		
CH$_3$	57.02	2.05[c]	9.55	56.52[c]	2.57	8.97
CH$_2$CH$_3$	57.1[c]		9.2			
CH(CH$_3$)$_2$	57.4		8.9	56.6		9.1
C(CH$_3$)$_3$	57.85	2.11	8.64	56.67	2.58	9.23
CH$_2$PO(OC$_2$H$_5$)$_2$	58.1[c]			55.1		
CN	60.06		10.95[e]	56.44	2.61	9.05
CH=NOH	58.1		9.6	57.2[d]		8.9
C(CH$_3$)=NOH	58.6			56.7		
SiH$_3$	49.5[d]		9.6[d]	54.7		11.1
SiCl$_3$	51.4[c]			55.7		
SH	60.1[c]					
SO$_3$CH$_3$	59.8		10.0	56.1[d]		8.7
Se—Se'—C$_6$H$_5$	59.29[c]		9.84	55.27[c]	2.59	8.80
P(C$_6$H$_5$)$_2$	55.01	2.11[e]		55.32		9.61[d]
H	55.95	2.46	10.01	55.95	2.46	10.01

Source. Reference 102.

[a]In hertz. The errors for the one-bond couplings are at least within ±0.1 Hz except for the AB satellite spectra, where only the two inner lines are observed; in these cases the errors are within ±0.5 Hz. For the long-range couplings the errors are within ±0.03 Hz in cases where these couplings are given with two decimals; in all other cases the errors are ±0.1 Hz.

Table 5.9. (Continued)

3, 4	1, 7	2, 7	3, 7	Other Couplings [b]
56.22				$\begin{cases} \text{F, 1: } -245.1; \text{F, 2: } +21.02 \\ \text{F, 3: } +7.79; \text{F, 4: } +3.20 \end{cases}$
56.08[d]				
56.11				
56.12[c]				
56.1				
56.15				1, 8: 2.33; 2, 8: 4.08
55.31				
56.0				
56.2				
56.10				2, 8: 2.89[c]
55.9				
55.8				
56.6[d]				
55.2				
55.1				
55.4[d]	74.9	+2.49	+4.59	4, 7: -1.04
55.2				
55.2				$\begin{cases} \text{F, 1: } +61.1; \text{F, 2: } +4.04 \\ \text{F, 3: } +1.38; \text{F, 4: } 1.10 \end{cases}$
56.25[c]	44.26	3.22[d]	3.83	
56.0	45.5			7, 8: 33.8
56.0	43.3			7, 8: 34.5
55.83	43.21[c]	1.95	3.18	2, 8: 2.23; 7, 8: 35.43
55.0				$\begin{cases} \text{P, 1: } -8.95; \text{P, 2: } +6.67 \\ \text{P, 3: } -2.93; \text{P, 4: } +3.49 \end{cases}$
55.10			5.51[c]	
54.7[c]	62.8			
	61.7			7, 8: 42.5
55.4[c]				Si, 1: 71.6; Si, 2: 4.88; Si, 3: 6.09
55.5				Si, 1: ~117; Si, 3: 9.1
56.0				
55.2				
56.47[c]				$\begin{cases} \text{Se, 1: } -122.2; \text{Se}', 1: 4.74 \\ \text{Se, 2: } +11.58; \text{Se, 3: } +1.81 \end{cases}$ P, 1: -12.46; P, 2: +19.63; P, 3: +6.82
55.95				

[b]Carbons in the side-chain are numbered consecutively.
[c]Only inner lines of AB spectrum are observed.
[d]Only three lines of AB spectrum are observed.
[e]Only lines around one of the carbon signals are observed.

thors on H–H, C–H, C–C, and C–N coupling constants in several polycyclic aromatic hydrocarbons and aromatic nitrogen heterocyclics.[41–80] These numbers are accompanied by values of those constants from FP–INDO calculations.[81,82] It should be recalled that since these calculations are done in the INDO approximation, the contribution of the π-electron system is included (see previous section).

The most careful comparisons of theory and experiment for spin–spin coupling in aromatic compounds have been for the one-ring compounds benzene,[83] **1**, and pyridine, **7**. Several detailed studies have been made of proton–proton coupling in six-membered aromatic nitrogen-containing molecules related to pyridine.[83–86] Molecules that have been studied are pyridine, the pyridinium anion, pyridine *n*-oxide, protonated pyridine *n*-oxide, aminopyridine, and chloropyridine. Other studies have been done of chalcogen heterocyclics,[87] furan,[84] thiophenes,[88] and selenophenes.[89] Studies have also been done of the nonalternant polycyclic aromatic hydrocarbon azulene[90] and a variety of compounds containing a single aromatic benzene ring.[91–93] There is also a considerable amount of work involving coupling between protons and fluorine nuclei.[95–99] Wasylichen and Schaefer[94] have presented calculated and experimental values for coupling constants between aromatic ring protons and protons in a substituent to the ring (benzylic interactions).

A study of carbon–carbon spin–spin coupling constants in monosubstituted benzenes has been published by Wray, Ernst, Lund, and Jakobsen[102] (see Table 5.9). They found a correlation between coupling constants of neighbors, and coupling constants of carbons separated by three bonds, with electronegativity of the first atom of the substituent. Other couplings did not depend strongly on

Table 5.10. Electron-Withdrawing Power, ΔQ_x, Calculated by INDO–MO Theory for the Group X in Molecules CH_3X Relative to CH_4

X ΔQ_x (electrons x 10^3)	F -221.5	OH -122.4	OCH_3 -122.6	NO_2 -135.3	NH_2 -50.0
$NHCH_3$ -49.9	$N(CH_3)_2$ -50.5	$NHNH_2$ -55.1	NCO -115.6	$COCH_3$ -10.6	
X ΔQ_x (electrons x 10^3)	CO_2CH_3 $+6.0$	CHO -6.5	CFO -27.7	CH_3 $+9.4$	CH_2CH_3 -16.0	$CH(CH_3)_2$ $+20.2$
$C(CH_3)_3$ $+22.9$	CN -59.8	CH=NOH -23.1	$C(CH_3)=NOH$ -9.3	H 0.0	

Source. Reference 102.

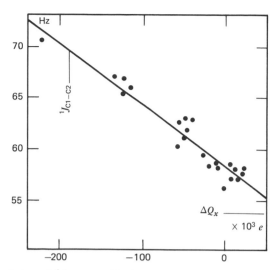

Figure 5.1. Correlation of $^1J_{C1-C2}$ with the group-electron-withdrawing power, ΔQ_x, of the substituent. (From reference 102.)

substituent change. They defined a new substituent parameter, called "group-electron-withdrawing power," which was defined for a substituent group, X, as the total charge density of the CH_3 group in CH_3X minus the total charge density of the CH_3 group in CH_4. The charge densities were calculated using the INDO-MO method. Correlation of experimental coupling constants with this parameter was better (see Table 5.10 and Figure 5.1) than with electronegativity.

Marshall, Faehl, and Kattner[103] have considered C–C coupling constants in a number of aromatic compounds. Berger and Zeller[104] have considered C–C coupling constants in phenanthrene derivatives. Hansen, Poulsen, and Berg[105] have measured C–C coupling constants in phenyl-substituted ethylene, naphthalene, phenanthrene, and cyclopentadienone. Their data suggest that the signs of coupling constants over more than two bonds alternate in aromatic molecules. Hansen and Berg[106] have also considered coupling constants in anthracene,9,10,di-hydroanthracene, and phthalic acid derivatives. Llinare, Faure, and Vincent[107] have studied the C–C coupling constants and derivatives of thiazole.

The effect of a substituent on the H–H coupling constants in benzene has been studied experimentally in some detail.[108–113] Some experimental results, along with calculations based on the FP-INDO method,[83] appear as Table 5.11. Agreement of theory with experiment is encouraging. In disubstituted benzenes, the effect of the substituents on H–H coupling appears to be approximately additive.[83]

A consideration of the experimental data presented in Tables 5.1 to 5.11 in conjunction with the theoretical predictions leads one to conclude that recent

Table 5.11. J_{HH} Values for Monosubstituted Benzenes

Substituent X	Calculated[a,b]					
	J_{12}	J_{13}	J_{14}	J_{15}	J_{23}	J_{24}
Li	6.29	' 3.49	2.56	-1.73	8.94	1.36
	(-1.86)	(1.36)	(1.41)	(-3.86)	(0.79)	(-0.77)
CH_3	7.78	2.27	1.27	1.89	8.30	2.02
	(0.37)	(0.14)	(0.12)	(-0.24)	(0.15)	(-0.11)
CHO^c	7.68	2.13	1.37	1.70	8.27	1.91
	(-0.47)	(0.00)	(0.22)	(-0.43)	(0.12)	(-0.22)
CCH	7.78	2.26	1.29	1.87	8.27	1.97
	(-0.37)	(0.13)	(0.14)	(-0.26)	(0.12)	(-0.16)
CH	7.73	2.21	1.29	1.84	8.21	1.91
	(-0.42)	(0.08)	(0.14)	(-0.29)	(0.06)	(-0.22)
NO	7.95	2.36	0.90	2.78	7.86	2.32
	(-0.19)	(0.23)	(-0.25)	(0.65)	(-0.29)	(0.19)
NH_2	8.37	2.04	1.03	2.64	8.15	2.25
	(0.22)	(-0.09)	(-0.12)	(0.51)	(0.00)	(0.12)
OH	8.69	1.87	0.92	2.92	7.97	2.44
	(0.54)	(-0.26)	(-0.23)	(0.79)	(-0.18)	(0.31)
OCH_3	8.83	1.96	0.94	2.96	7.96	2.49
	(0.68)	(-0.17)	(-0.21)	(0.83)	(-0.19)	(0.36)
$NO_2{}^d$	8.02	1.81	1.01	2.53	7.79	2.20
	(-0.13)	(-0.32)	(-0.14)	(0.40)	(-0.36)	(0.07)
F	9.04	1.69	0.82	3.26	7.75	2.62
	(0.89)	(-0.44)	(-0.33)	(1.13)	(-0.40)	(0.49)
$N + (CH_3)_3$	8.45	1.66	0.82	3.22	7.18	2.20
	(0.30)	(-0.47)	(-0.33)	(1.09)	(-0.97)	(0.07)

Source. Reference 83.

[a]Values in Hz.

[b]Values in parentheses are the differences between particular J_{HH} values and the corresponding couplings in benzene. Calculated J_{HH} values are referenced to calculated benzene results; experimental J_{HH} values are referenced to experimental J_{HH} data.

[c]For a molecular geometry with the HCO plane perpendicular to the benzene ring, the

Table 5.11. (Continued)

Experimental						
J_{12}	J_{13}	J_{14}	J_{15}	J_{23}	J_{24}	Reference
6.73	1.54	0.77	0.74	7.42	1.29	e
(−0.81)	(0.17)	(0.08)	(−0.63)	(0.12)	(−0.08)	
7.64	1.25	0.60	1.87	7.53	1.51	109
(0.10)	(−0.12)	(−0.09)	(0.50)	(−0.01)	(0.14)	
7.71	1.35	0.62	1.75	7.48	1.25	e
(0.17)	(−0.02)	(−0.07)	(0.38)	(−0.06)	(−0.12)	
7.72	1.32	0.64	1.77	7.61	1.37	110
(0.16)	(−0.06)	(−0.04)	(0.39)	(0.05)	(−0.01)	
7.79	1.28	0.63	1.76	7.68	1.30	111
(0.25)	(−0.09)	(−0.06)	(0.39)	(0.14)	(−0.07)	
7.91	1.28	0.56	1.97	7.41	1.36	112
(0.35)	(−0.10)	(−0.12)	(0.59)	(0.15)	(−0.02)	
8.02	1.11	0.47	2.53	7.39	1.60	112
(0.48)	(−0.26)	(−0.22)	(1.16)	(−0.15)	(0.23)	
8.17	1.09	0.49	2.71	7.40	1.74	112
(0.63)	(−0.28)	(−0.20)	(1.34)	(−0.14)	(0.37)	
8.30	1.03	0.44	2.74	7.36	1.76	112
(0.76)	(−0.34)	(−0.25)	(1.37)	(−0.18)	(0.39)	
8.36	1.18	0.55	2.40	7.47	1.48	112
(0.82)	(−0.19)	(−0.14)	(1.03)	(−0.07)	(0.11)	
8.36	1.07	0.43	2.74	7.47	1.83	113
(0.82)	(−0.30)	(−0.26)	(1.37)	(−0.07)	(0.45)	
8.55	0.92	0.48	3.05	7.46	1.70	112
(1.01)	(−0.45)	(−0.21)	(1.68)	(−0.08)	(0.33)	

values 7.56, 2.28. 1.32, 1.73, 8.25, and 1.79 were obtained for $J_{12}, J_{13}, J_{14}, J_{15}, J_{23}$, and J_{24}, respectively.

[d]For a molecular geometry with the NO_2 plane perpendicular to the benzene ring, the values 7.92, 1.89, 0.99, and 2.44 were obtained for J_{12}, J_{13}, J_{14}, and J_{15}, respectively.

[e]S. Castellano, private communications.

theories of spin–spin coupling, particularly the FP method using the INDO approximation, lead to fairly plausible results for aromatic compounds, at least when compared to MO theories of other parameters in large molecules. Agreement seems to be better for H–H than for C–H and C–C coupling.

REFERENCES

1. H. S. Gutowsky and D. W. McCall, *Phys. Rev.* **82**, 748 (1951).

2. H. S. Gutowsky, D. W. McCall, and C. P. Slichter, *Phys. Rev.* **84**, 589 (1951).

3. E. L. Hahn and D. E. Maxwell, *Phys. Rev.* **84**, 1246 (1951).

4. E. L. Hahn and D. E. Maxwell, *Phys. Rev.* **88**, 1070 (1952).

5. N. F. Ramsey, *Phys. Rev.* **91**, 303 (1953).

6. M. Barfield and D. M. Grant *in* J. S. Waugh (Ed.): *Advances in Magnetic Resonace*, Vol. 1, Academic, New York, 1965.

7. J. N. Murrell, *Prog. NMR Spectrosc.* **6**, 1 (1971).

8. J. Kowalewski, *Prog. NMR Spectrosc.* **11**, 1 (1977).

9. J. D. Memory, *Quantum Theory of Magnetic Resonance Parameters*, McGraw-Hill, New York, 1968.

10. H. M. McConnell, *J. Chem. Phys.* **24**, 460 (1956).

11. J. A. Pople and D. P. Santry, *Mol. Phys.* **8**, 1 (1964).

12. R. Hoffman, *J. Chem. Phys.* **39**, 1397 (1963).

13. J. A. Pople and D. P. Santry, *Mol. Phys.* **8**, 269 (1964).

14. C. C. J. Roothaan, *Rev. Mod. Phys.* **23**, 69 (1951).

15. J. A. Pople, D. P. Santry, and G. A. Segal, *J. Chem. Phys.* **43**, S129 (1965).

16. J. A. Pople and G. A. Segal, *J. Chem. Phys.* **43**, S136 (1965).

17. J. A. Pople and G. A. Segal, *J. Chem. Phys.* **44**, 3289 (1966).

18. J. A. Pople and D. L. Beveridge, *Approximate Molecular Orbital Theory*, McGraw-Hill, New York, 1970.

19. J. A. Pople, D. L. Beveridge, and P. A. Dobosh, *J. Chem. Phys.* **46**, 2026 (1967).

20. E. A. G. Armour and A. J. Stone, *Proc. R. Soc.* **A302**, 25 (1967).

21. R. Ditchfield and J. N. Murrell, *Mol. Phys.* **14**, 481 (1968).

22. R. Ditchfield, *Mol. Phys.* **17**, 33 (1969).

23. J. A. Pople, J. W. McIver, and N. S. Ostlund, *Chem. Phys. Lett.* **1**, 465 (1967).

24. J. A. Pople, J. W. McIver, and N. S. Ostlund, *J. Chem. Phys.* **49**, 2960 (1968).

25. J. A. Pople, J. W. McIver, and N. S. Ostlund, *J. Chem. Phys.* **49**, 2965 (1968).

26. N. S. Ostlund, M. D. Newton, J. W. McIver, and J. A. Pople, *J. Magn. Resonance* **1**, 298 (1969).

27. A. C. Blizzard and D. P. Santry, *Chem. Commun.* **1970**, 87 (1970).

28. A. C. Blizzard and D. P. Santry, *J. Chem. Phys.* **55**, 950 (1971). Erratum: *J. Chem. Phys.* **58**, 4714 (1973).

29. R. Ditchfield, N. S. Ostlund, J. N. Murrell, and M. A. Turpin, *Mol. Phys.* **18**, 433 (1970).

30. H. Nakatsuji, *J. Chem. Phys.* **61**, 3728 (1974).

31. E. Hiroike, *J. Phys. Soc. Japan* **22**, 379 (1967).

32. E. Hiroike, *J. Phys. Soc. Japan* **23**, 1079 (1967).

33. C. Barbier, D. Gagnaire, G. Berthier, and B. Levy, *J. Magn. Resonance* **5**, 11 (1971).

34. A. Denis and J. P. Malrieu, *Mol. Phys.* **23**, 581 (1972).

35. H. M. McConnell, *J. Mol. Spectrosc.* **1**, 11 (1957).

36. W. J. Van der Hart, *Mol. Phys.* **20**, 399 (1971).

37. P. V. Schastnev, N. D. Chuvylkin, and G. M. Zhidomirov, *Theor. Eksp. Khim.* **7**, 86 (1971).

38. M. Barfield, *J. Chem. Phys.* **48**, 4458 (1968).

39. M. Barfield, *J. Chem. Phys.* **49**, 2145 (1968). Erratum: *J. Chem. Phys.* **51**, 2291 (1969).

40. R. A. Hoffman, *Mol. Phys.* **1**, 326 (1958).

41. D. G. De Kowalewski, R. Buitrago, and R. Yommi, *J. Mol. Struc.* **11**, 195 (1972).

42. H. B. Evans, Jr., A. R. Tarpley, and J. H. Goldstein, *J. Phys. Chem.* **72**, 2552 (1968).

43. J. M. Read, R. E. Mayo, and J. H. Goldstein, *J. Mol. Spectrosc.* **22**, 419 (1967).

44. S. Castellano and R. Kostelnik, *Tetrahedron Lett.* **1967**, 5211 (1967).

45. J. M. Read, R. E. Mayo, and J. H. Goldstein, *J. Mol. Spectrosc.* **21**, 235 (1966).

46. N. K. Wilson and J. B. Stothers, *J. Magn. Resonance* **15**, 31 (1974).

47. Y. Sasaki and M. Suzuki, *Chem. Pharm. Bull. (Japan)* **17**, 1090 (1969).

48. N. Jonathan, S. Gordon, and B. P. Dailey, *J. Chem. Phys.* **36**, 2443 (1962).

49. B. Dischler and G. Englert, *Z. Naturforsch. A* **16**, 1180 (1961).

50. M. W. Jarvis and A. G. Moritz, *Aust. J. Chem.* **24**, 89 (1971).

51. J. B. Stothers, C. T. Tan, and N. K. Wilson, *Org. Magn. Resonance* **9**, 408 (1977).

52. J. D. Memory, G. W. Parker, and J. C. Halsey, *J. Chem. Phys.* **45**, 3567 (1966).

53. P. E. Hansen and A. Berg, *Acta Chem. Scand.* **25**, 3377 (1971).

54. A. R. Tarpley, Jr. and J. H. Goldstein, *J. Phys. Chem.* **76**, 515 (1972).

55. F. J. Weigert and J. D. Roberts, *J. Am. Chem. Soc.* **89**, 2967 (1967).

56. N. Muller and D. E. Pritchard, *J. Chem. Phys.* **31**, 758, 1471 (1959).

57. P. Granger and M. Maugras, *Chem. Phys. Lett.* **24**, 331 (1974).

58. P. C. Lauterbur, *J. Am. Chem. Soc.* **83**, 1838 (1961).

59. F. J. Weigert and J. D. Roberts, *J. Am. Chem. Soc.* **94**, 6021 (1972). H. J. Bernstein, private communication.

60. P. E. Hansen, O. K. Poulsen, and A. Berg, *Org. Magn. Resonance* **7**, 475 (1975).

61. M. Hansen and H. J. Jakobsen, *J. Magn. Resonance* **10**, 74 (1973).

62. R. L. Lichter and J. D. Roberts, *J. Am. Chem. Soc.* **93**, 5218 (1971).

63. J. P. Dorie, M. L. Martin, S. Barnier, M. Blain, and S. Odiot, *Org. Magn. Resonance* **3**, 661 (1971).

64. J. P. Jacobsen, O. Snerling, E. J. Pedersen, J. T. Nielsen, and K. Schaumburg, *J. Magn. Resonance* **10**, 130 (1973).

65. V. M. S. Gil and A. J. L. Pinto, *Mol. Phys.* **16**, 623 (1969).

66. K. Tori and M. Ogata, *Chem. Pharm. Bull. (Japan)* **12**, 272 (1964).

67. G. S. Reddy, R. T. Hobgood, Jr., and J. H. Goldstein, *J. Am. Chem. Soc.* **84**, 336 (1962).

68. P. J. Black and M. L. Heffernan, *Aust. J. Chem.* 17, 558 (1964).

69. P. A. Claret and A. G. Osborne, *Spectrosc. Lett.* 6, 103 (1973).

70. A. R. Katritzky and Y. Jakeuchi, *J. Chem. Soc. Perkin Trans.* 2 11, 1682 (1972).

71. P. J. Black and M. L. Heffernan, *Aust. J. Chem.* 18, 707 (1965).

72. H. H. Limbach, W. Seiffert, E. Ohmes, and H. Zimmermann, *Ber. Bunsenges. Phys. Chem.* 74, 617 (1970).

73. P. J. Black and M. L. Heffernan, *Aust. J. Chem.* 19, 1287 (1966).

74. J. P. Kokko and J. H. Goldstein, *Spectrochim. Acta* 19, 1119 (1963).

75. G. Miyajima, K. Takahashi, and H. Sugiyama, *Org. Magn. Resonance* 6, 181 (1974).

76. F. J. Weigert, J. Husan, and J. D. Roberts, *J. Org. Chem.* 38, 1313 (1973).

77. P. C. Lauterbur, *J. Chem. Phys.* 43, 360 (1965).

78. K. Tori and J. Nakagawa, *J. Phys. Chem.* 68, 3163 (1964).

79. E. R. Malinowski, L. Z. Pollara, and J. P. Larmann, *J. Am. Chem. Soc.* 84, 2649 (1962).

80. J. D. Memory, unpublished data.

81. S. A. T. Long and J. D. Memory, *J. Magn. Resonance* 29, 119 (1978).

82. S. A. T. Long and J. D. Memory, *J. Magn. Resonance* 44, 355 (1981).

83. G. E. Maciel, J. W. McIver, N. S. Ostlund, and J. A. Pople, *J. Am. Chem. Soc.* 92, 4506 (1970).

84. G. E. Maciel, J. W. McIver, N. S. Ostlund, and J. A. Pople, *J. Am. Chem. Soc.* 92, 4497 (1970).

85. L. Ernst, D. N. Lincoln, and V. Wray, *J. Magn. Resonance* 21, 115 (1976).

86. R. H. Contreras and V. J. Kowalewski, *J. Magn. Resonance* 39, 291 (1980).

87. V. Galasso, *Chem. Phys. Lett.* 32, 108 (1975).

88. V. Galasso, *Chem. Phys. Lett.* 21, 54 (1973).

89. V. Galasso and A. Bigotto, *Org. Magn. Resonance* 6, 475 (1974).

90. D. J. Bertelli, T. G. Andrews, and P. O. Crews, *J. Am. Phys. Soc.* 91, 5286 (1969).

91. M. A. Cooper and S. L. Mavatt, *J. Am. Chem. Soc.* 92, 1605 (1970).

92. C. S. Cheung, M. A. Cooper, and S. L. Mavatt, *Tetrahedron* 27, 701 (1971).

93. V. Galasso, G. Pellizer, A. Lisini, and A. Bigotto, *Org. Magn. Resonance* 7, 591 (1975).

94. R. Wasylichen and T. Schaefer, *Can. J. Chem.* 50, 1852 (1972).

95. M. S. Gopinathan and P. T. Narasimhan, *Mol. Phys.* 22, 543 (1971).

96. K. Schaumberg, *J. Magn. Resonance* 3, 360 (1970).

97. H. Bildsoe and K. Schaumberg, *J. Magn. Resonance* 14, 223 (1974).

98. H. Bildsoe and K. Schaumberg, *J. Magn. Resonance* 13, 255 (1974).

99. I. Brown and D. W. Davies, *Chem. Commun.* 1972, 939 (1972).

100. R. L. Lichter and J. D. Roberts, *Spectrochim. Acta* A26, 1813 (1970).

101. F. J. Weigert and J. D. Roberts, *J. Am. Chem. Soc.* 94, 6021 (1972).

102. V. Wray, L. Ernst, T. Lund, and H. J. Jakobsen, *J. Magn. Resonance* 40, 55 (1980).

103. J. L. Marshall, L. G. Faehl, and R. Kattner, *Org. Magn. Resonance* 12, 163 (1979).

104. S. Berger and K. Zeller, *Org. Magn. Resonance* 11, 303 (1978).

105. P. E. Hansen, O. K. Poulsen, and A. Berg, *Org. Magn. Resonance* 12, 43 (1979).

106. P. Hansen and A. Berg, *Org. Magn. Resonance* 12, 50 (1979).

107. J. Llinarès, R. Faure, and É.-J. Vincent, *C. R.. Acad. Sci. Paris*, **C289**, 133 (1979).

108. J. M. Read, Jr., R. E. Mayo, and J. H. Goldstein, *J. Mol. Spectrosc.* **21**, 235 (1966).

109. M. P. Williamson, R. J. Kostelnik, and S. M. Castellano, *J. Chem. Phys.* **49**, 2218 (1968).

110. S. Castellano and J. Lorenc, *J. Phys. Chem.* **69**, 3552 (1965).

111. K. Hayamizu and O. Yamamoto, *J. Mol. Spectrosc.* **25**, 422 (1968).

112. S. Castellano, C. Sun, and R. Kostelnik, *Tetrahedron Lett.* **51**, 5205 (1967).

113. S. Castellano, R. Kostelnik, and C. Sun, *Tetrahedron Lett.* **46**, 4635 (1967).

Six

Relaxation Times and Other Time-Dependent Phenomena

Within a few years after the first observations of NMR chemical shifts in organic molecules were made in the late 1940s, NMR techniques were used to measure exchange rates.[1,2] Since then, the great versatility and utility of NMR to measure rates in the range of about 10^{-2} to 10^{10} sec^{-1} and to examine various aspects of molecular motion have been firmly established. In this chapter, we review briefly the principles of and the resulting methods for the high-resolution NMR study of dynamic processes involving aromatic organic molecules. An extensive body of literature on aspects of dynamic NMR employing proton spectroscopy exists and has been covered thoroughly in several texts and review articles,[3-6] including some of the basic texts given in the reference section of Chapter 1. Thus we here illustrate these principles and methods with examples from the recent literature, emphasizing pulsed NMR, relaxation times, and nuclei other than protons.

Comprehensive reviews of many aspects of time-dependent phenomena observable by NMR spectroscopy are available. These include reviews on dynamic NMR in general,[3,5,7-9] on dynamic ^{13}C NMR,[10-14] on relaxation times and nuclear Overhauser enhancements (NOEs),[15-21] diffusion and chemical exchange,[15,22,23] line shape analysis,[7,24,25] CIDNP,[26] reaction mechanisms and intermediates,[27] carbonium ion rearrangements,[28] proton transfer,[29] flow and stopped-flow reactions,[30] and specific applications to polycyclic aromatic compounds employing ^{13}C NMR.[31]

RELAXATION TIMES

Measurements of dynamic processes in the early years of NMR spectroscopy were generally based on changes in line shapes and line widths. The observable rate range was therefore restricted to approximately 10^1 to 10^6 sec^{-1}. With measurements of spin-lattice and spin-spin relaxation times, the observable range can be extended by several orders of magnitude in both directions, to approximately 10^{-2} to 10^{10} sec^{-1}. Accompanying this extension of the observable rate range are some remarkable bonuses in the form of information about molecular mobility and anisotropic molecular motion. ^{13}C spin-lattice relaxation times have proved especially valuable in the latter regard, since they are more readily interpreted than are proton relaxation times.

The two characteristic times—T_1 for spin-lattice, or longitudinal, relaxation and T_2 for spin-spin, or transverse, relaxation—describe different time-dependent processes in the nuclear spin system.[32,33] Both T_1 and T_2 processes involve non-radiative transitions. Interactions that involve the transfer of excess nuclear spin energy to other degrees of freedom of the molecular system in which the spins are embedded (the lattice) reestablish thermal equilibrium between the spin system and the lattice with the characteristic time T_1. The spin-lattice relaxation time thus describes the rate at which the thermal distribution of spins among the nuclear energy levels is reestablished after a perturbing event. Transitions between nuclear spin levels, and thus contributions to T_1, may arise from any local fluctuating magnetic fields with frequency components at the nuclear Larmor frequency. Dynamic occurrences in the sample such as molecular rotations, with typical correlation times of 10^{-10} to 10^{-13} sec, produce such local field variations with time. Thus, T_1 values reflect both the amount and nature of molecular motion.

Mutual exchanges of spin energy between neighboring nuclei may produce transitions between nuclear spin states without affecting the thermal distribution of spins among the states. Thus, during these exchanges, the overall energy of the spin system is conserved. The spin-spin relaxation process governs the lifetime of a given spin state, and hence affects the shape of a given signal. Low-frequency processes such as chemical exchange can contribute substantially to spin-spin relaxation and its characteristic T_2 values.

Spin-Lattice Relaxation

There are several phenomena that can contribute to spin-lattice relaxation—internuclear dipole-dipole interactions, spin rotation in small or very symmetrical molecules, electric quadrupolar interactions, chemical shift anisotropies (especially at high fields), scalar coupling interactions, and electron-nuclear interactions in the presence of paramagnetic materials. The relaxation mech-

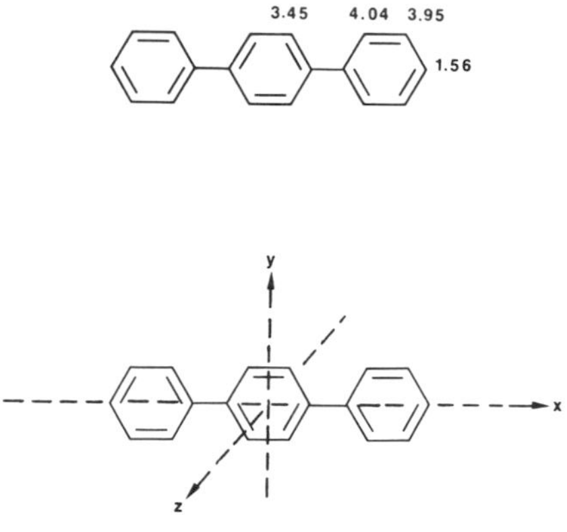

Figure 6.1. ^{13}C spin–lattice relaxation times, in seconds, for the protonated carbons, and the molecular rotational axes for p-terphenyl. (From reference 34.)

anisms attributable to these phenomena have been reviewed in detail in several texts[3,10] and articles.[15,18,21]

Of the spin–lattice relaxation mechanisms, the most important for typical medium-sized molecules in solution is internuclear dipole–dipole relaxation. Dipolar relaxation depends upon the rate of change of the internuclear vector between two magnetic nuclei with respect to the applied magnetic field vector. Motion at or near the Larmor frequency is most efficient in producing dipolar relaxation. Therefore, in nonviscous liquids, a slow molecular motion is usually more efficient than a fast one. For example, in p-terphenyl, slow end-over-end tumbling perpendicular to the long molecular axis relative to faster rotation about this axis shortens the observed T_1 values for CH carbons on the long axis relative to CH carbons off this axis, as is shown in Figure 6.1.

The relaxation of carbons having directly bonded protons is generally dominated by dipolar interactions with those protons. In these cases, the spin–lattice relaxation time is given by[35,36]

$$\frac{1}{T_1} = N\hbar^2 \gamma_C^2 \gamma_H^2 r^{-6} \tau_c \tag{6.1}$$

where N is the number of directly bonded protons, γ_C and γ_H are the magnetogyric ratios of ^{13}C and ^{1}H, r is the C–H internuclear distance, and τ_c is the effective correlation time for rotational reorientation. Equation 6.1 is valid in the motional narrowing approximation—that is, $1/\tau_c$ must be much greater than the ^{13}C and ^{1}H resonance frequencies. For carbons without directly bonded protons, other relaxation mechanisms may be important, or even predominant. In general, the spin–lattice relaxation time can be expressed in terms of the T_1 contributions from various mechanisms:

$$\frac{1}{T_1} = \frac{1}{T_1^{dd}} + \frac{1}{T_1^{sr}} + \frac{1}{T_1^{csa}} + \frac{1}{T_1^{other}} \tag{6.2}$$

where T_1^{dd} is the dipolar contribution, T_1^{sr} the spin–rotation contribution, T_1^{csa} the chemical shift anisotropy contribution, and T_1^{other} the contribution from other mechanisms.

An important result of the internuclear dipole–dipole interaction is the nuclear Overhauser effect. If heteronuclei, X, that are coupled to the observed nuclei and contribute to the relaxation of the observed nuclei through dipolar processes are irradiated at their resonance frequencies at an rf level sufficient to saturate the X transitions, the resonance intensities of the observed nuclei, A, are changed by a factor of $1 + \gamma_X/\gamma_A$, where γ_X and γ_A are the magnetogyric ratios of X and A, respectively. The nuclear Overhauser enhancement (NOE) is the difference in intensities with and without decoupling of X: $\eta = \gamma_X/2\gamma_A$. If the relaxation of A is not completely dipolar, the NOE is proportionately less:

$$\eta = \frac{1}{2} \frac{\gamma_X}{\gamma_A} \frac{R_1^{dd}}{R_1} \tag{6.3}$$

where R_1^{dd} is the spin–lattice relaxation rate of A caused by dipole–dipole inter-actions with X and R_1 is the total spin–lattice relaxation rate of A. The relaxa-tion rate R_1 is defined as $R_1 = 1/T_1$.

When the motional narrowing assumption is not valid, or when heteronuclei coupled to A are irradiated selectively, the NOE is more complicated.[37] How-ever, for typical proton broad-band-decoupled spectra of ^{13}C, the NOE can lead to signal/noise enhancement of the CH resonances by a factor of approx-imately 3. This factor is different, of course, for observed nuclei having a differ-ent magnetogyric ratio. ^{15}N, for example, has a large negative magnetogyric ratio. Depending on the contribution of ^{1}H-^{15}N dipolar relaxation to the total T_1 of the ^{15}N of interest, the ^{15}N signal intensity, $1 + \eta$, may be small and positive, negative, or even zero in the presence of ^{1}H decoupling.

^{13}C spin–lattice relaxation in several condensed aromatic compounds—naphthalene, **1**, phenanthrene, **2**, pyrene, **3**, acenaphthene, **4**, and xanthone, **5**—

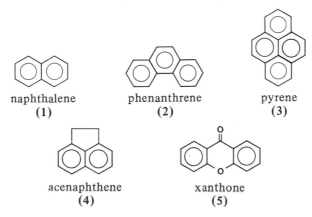

naphthalene (1)   phenanthrene (2)   pyrene (3)   acenaphthene (4)   xanthone (5)

has recently been examined in detail by Alger and his co-workers.[38] By measuring T_1 and NOE values of the same carbons at two different resonance frequencies, these workers were able to determine the relative contributions of the dipolar and chemical shift anisotropy mechanisms to the spin–lattice relaxation. For ^{13}C nuclei having attached protons, the dipolar mechanism dominated the relaxation, as evidenced by NOE values $(1 + \eta)$ ranging from 2.7 to 3.0, at both 25.1 and 75.3 MHz. The corresponding T_1 values for these CH carbons range from 3 to 11 sec. In contrast to that of the protonated carbons, the relaxation of the quaternary carbons was dominated by chemical shift anisotropy at high frequency (75.3 MHz), where the average quaternary carbon NOE was 1.5 and T_1^{csa} was about 30–90 sec. At the lower frequency (25.1 MHz), the chemical shift anisotropy and dipolar mechanisms contributed approximately equally to the nonprotonated carbon relaxation. Values of the NOE, T_1^{dd}, and T_1^{csa} at 25.1 MHz were approximately 2.1 sec, more than 100 sec, and 300–1,000 sec, respectively.

Other studies of ^{13}C spin–lattice relaxation times in polycyclic aromatic compounds have been reported. These earlier studies all support the predominance of dipole–dipole interactions in the relaxation of hydrogen-bearing carbons in these compounds.[39-45] An example of a carbon spectrum that shows clearly the differences in signal intensities of protonated and nonprotonated carbons that result from the smaller NOEs and longer T_1 values of the latter carbons is given in Figure 6.2.

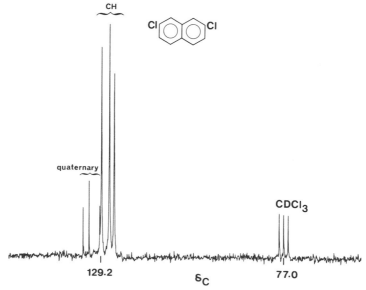

Figure 6.2. The proton noise-decoupled ^{13}C NMR spectrum of 2,7-dichloronaphthalene. (From reference 46.)

Because of their dependence on the correlation times for molecular motion, spin–lattice relaxation times normally show a temperature dependence. For ^1H, ^{19}F, and other abundant nuclei subject to several different relaxation processes, the temperature dependence is often difficult to interpret. If the relaxation is wholly dipole–dipole, however, as it often is for ^{13}C nuclei, then the T_1 temperature dependence can be related to motional barriers through the usual relationships for a thermally activated process

$$\tau_c = \tau_c^0 \exp\left(E_a/RT\right) \tag{6.4}$$

where E_a is the activation energy, R the gas constant, and T the temperature. The slope of a plot of $1/T_1$ versus $1/T$ will provide a value of E_a. Measurements of the temperature dependences of spin–lattice relaxation times to provide activation energies have been reviewed recently.[11-12a] Most of these applications have involved the determination of barriers to methyl group rotation.[47-52]

Spin–Spin Relaxation

The spin–spin relaxation time, T_2, which is characteristic of the rate of exchange of spin energy between neighboring nuclei, is related to the natural width of a Lorentzian line by $T_2 = 1/\pi\Delta\nu_0$, where $\Delta\nu_0$ is the full line width at half height. This natural line width is generally quite small. However, any factor that effectively varies the relative energies of the spin levels and therefore increases the spread of nuclear precession frequencies will decrease the apparent T_2. Thus the observed line width, including the effects of field inhomogeneities, is characterized by an apparent spin–spin relaxation time $T_2^* = 1/\pi\Delta\nu$, where $\Delta\nu$ is the observed line width.

Chemical exchange and other low-frequency processes contribute to T_2 values, but not to T_1 values, which are affected primarily by high-frequency interactions near the Larmor frequency. The observed spin–spin relaxation time is given by

$$\frac{1}{T_2^*} = \frac{1}{T_2} + \frac{1}{T_2'} + \frac{\gamma\Delta H_0}{2} \tag{6.5}$$

where ΔH_0 is the applied magnetic field inhomogeneity, γ is the nuclear magnetogyric ratio, and T_2' includes contributions to the spin–spin relaxation time of exchange and other low-frequency processes. Consequently, line shape analysis permits the investigation of chemical exchange and the determination of kinetic data for relatively slow processes that are difficult to study by other methods.

Direct pulsed NMR measurements of T_2 values in dynamic systems are not so numerous as are those of T_1 values, since the experimental difficulties are much greater for T_2 measurements. In most nonviscous liquids, $T_{1\rho}$ measurements can be used to determine T_2, since in these situations $T_{1\rho} = T_2$.

In general, $T_2 \leqslant T_1$. For ^{13}C nuclei, T_2 is generally shorter for those carbons bonded to quadrupolar nuclei such as chlorine than for those bonded to non-quadrupolar nuclei such as hydrogen. An example is *o*-dichlorobenzene, **6**. The

o-dichlorobenzene

(6)

T_1 and T_2 values are essentially the same for the protonated carbons in **6**, about 7.7 and 6.4 sec. The measured T_2 for the chlorinated carbons is 4.2 sec, whereas T_1 for these carbons is 66 sec.[53] The short T_2 for the chlorine-bearing carbons arises from the low-frequency-modulated scalar interaction between the carbon and chlorine. Another situation that reduces T_2 but not T_1 is very fast relaxation of protons, which shortens the T_2 of carbons coupled to these protons so that $T_2 \ll T_1$.[54]

Older measurements of T_2 in exchanging systems have largely employed spin-echo techniques. The basic methods are summarized in Chapter 7, and extensive reviews of these applications have been published.[20,23,55] The later Fourier transform techniques have not been applied as extensively, but show great promise, especially for ^{13}C NMR studies. Several chapters in the text on dynamic NMR by Jackman and Cotton[5] are especially helpful in providing guides to T_2 measurements on dynamic NMR systems.

Applications of Relaxation Time Measurements

Spin–Lattice Relaxation Times and Molecular Motion

The dependence of spin–lattice relaxation times on the correlation times for molecular motion exemplified in Eq. 6.1 often allows one to gain useful insights into molecular mobility, anisotropic tumbling, and multiple intramolecular motions. Physical chemical applications of ^{13}C relaxation time measurements, such as studies of molecular diffusion and rotation, have been reviewed recently.[15] Similar applications of relaxation of ^{1}H and other nuclei have been summarized in several texts and articles.[3,4,17,23] Additionally, the utility of ^{13}C spin–lattice relaxation measurements in assessing anisotropic rotation of molecules has been reviewed recently.[12,18]

In monosubstituted benzenes, the spin–lattice relaxation of the protonated carbons is dominated by the dipolar mechanism.[56,57] Although the relaxation data in Figure 6.3 were obtained in different solvents, and thus precise inter-molecular comparisons cannot be made, motional anisotropy is clearly evidenced by the faster relaxation of carbons para to the substituent relative to that of

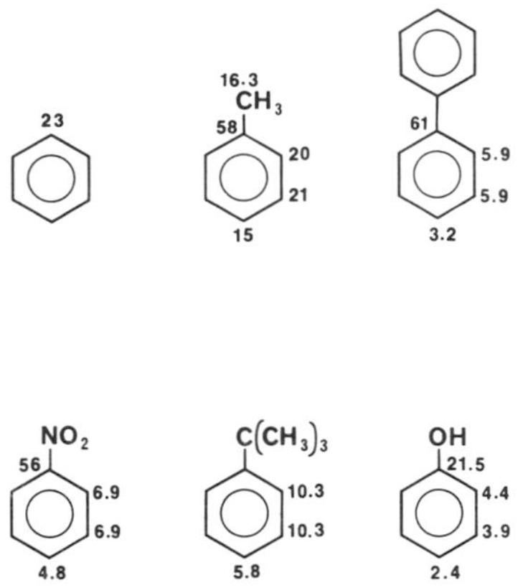

Figure 6.3. ^{13}C spin–lattice relaxation times, in seconds, for some substituted benzenes. The data were obtained at 25.2 MHz and ambient temperature. Benzene, toluene, and *t*-butylbenzene were examined as neat liquids, nitrobenzene and biphenyl as 20% solutions in acetone-d_6, and phenol as a 17 mole % solution in carbon tetrachloride. (From references 56 and 57.)

ortho and meta carbons. Preferred rotation about the C_2 symmetry axis does not shorten τ_c for the para carbons. Rather, it leads to less effective relaxation of the ortho and meta carbons because the orientations of their C–H bonds with respect to the static magnetic field change more rapidly than that of the para C–H bond. The ratio of tumbling about the long axis relative to the other molecular axes can be estimated from the dipolar T_1 values. For *t*-butylbenzene, this tumbling ratio is about 3.5.[57]

The effects of anisotropic molecular tumbling on the T_1 values of the protonated carbons in *p*-terphenyl were mentioned earlier in this chapter (see Figure 6.1). An additional feature apparent in these relaxation times is the smaller T_1 of hydrogen-bearing carbons on the central ring. The rings of *p*-terphenyl should be relatively free to rotate independently about the long molecular axis. The smaller T_1 values of carbons in the 2′, 3′, 5′, and 6′ positions of the central ring suggest that the ortho–ortho′ steric interactions of the protons on the central and terminal rings hinder slightly the rotation about the x axis of the central ring relative to the terminal rings, so that spin-lattice relaxation of the central carbons is somewhat more efficient.[34]

In phenalenone,[58] 7, protonated carbons on a twofold axis (C-2, C-5, and

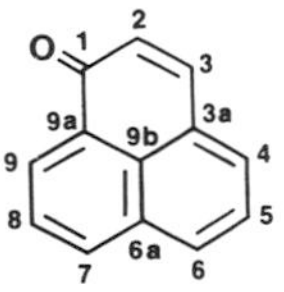

phenalenone
(7)

C-8) exhibit T_1 values shorter than those of carbons not on such an axis (C-3, C-6, C-9, and C-4)—1.4–1.5 sec compared to 1.9-2.0 sec. The molecular rotation in **7** thus appears to be predominantly about these axes. The quaternary carbons in this compound have long relaxation times, as expected. Carbons 3a, 6a, and 9a have T_1 values of 39–40 sec, indicating the inefficient dipolar relaxation of these carbons. The interior quaternary carbon, C-9b, has a T_1 of 71 sec. Based on the results of Alger and his co-workers[38] discussed earlier in this chapter, we might interpret this very long T_1 as indicating an important, perhaps even dominant, chemical shift anisotropy relaxation mechanism for C-9b.

Wasylishen and his co-workers[45] have examined the hydrodynamic rotation of triphenylene, **8**, employing ^1H, ^2H, and ^{13}C NMR relaxation time measurements. A correlation time for molecular reorientation, $\tau_c = 20.5 \times 10^{-12}$ sec, was obtained from the T_1 values for C-1 and C-2 in triphenylene, both 2.15 sec; thus the molecule appears to rotate freely in its molecular plane.

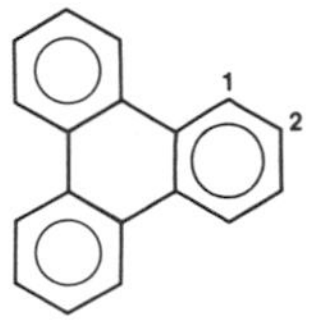

triphenylene
(8)

Investigations of anisotropic molecular rotation of triptycene, **9**, and fluorene, **10**, employing ^{13}C T_1 measurements and bond lengths derived from infrared

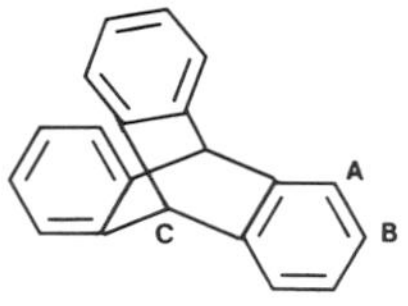

triptycene   fluorene
(9)   (10)

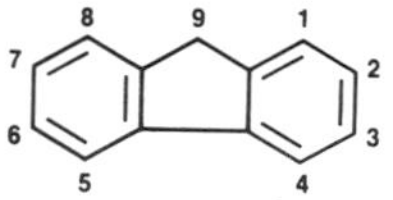

spectra were carried out by Harris and Newman.[59] Relaxation was dominated by dipole–dipole interactions for all the hydrogen-substituted carbons in both **9** and **10**. Two correlation times fully describe the rotational diffusion of **9**—$\tau_\parallel$ and $\tau_\perp$, for reorientation about the symmetry axis and reorientation of the symmetry axis, respectively. The results of the analysis of the T_1 data gave $\tau_\parallel = 48 \times 10^{-12}$ sec and $\tau_\perp = 16 \times 10^{-12}$ sec, with corresponding T_1 values of 1.95 sec at carbons A, 2.81 sec at carbons B, and 2.92 sec at the methine carbons C. Thus reorientation of the triptycene molecule about its symmetry axis is roughly three times slower than reorientation of that axis.

Fluorene, **10**, has C_{2v} symmetry, thus requiring three independent correlation times to fully describe its motion—τ_x, τ_y, and τ_z—for the motion about the axes

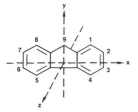

Figure 6.4. Molecular rotational axes[59] for fluorene, 10.

shown in Figure 6.4. From the measured T_1 values, values of T_1^{dd} were calculated for the several carbons in **10**, using a treatment derived by Woessner.[60] From relaxation times, T_1, of 6.04 sec for C-1, 4.46 sec for C-2, 5.17 sec for C-3, 6.03 sec for C-4, and 2.57 sec for C-9, the best fit correlation times were $\tau_x = 8 \times 10^{-12}$, $\tau_y = 23 \times 10^{-12}$, and $\tau_z = 6 \times 10^{-12}$ sec. These data, interpreted in terms of current hydrodynamic theories of molecular motion, suggest that fluorene behaves as an approximate ellipsoid in solution.

Similar relaxation studies[61] of ^{13}C and ^{14}N were made as functions of temperature for three neat liquid diazabenzenes—pyrimidine, **11**, pyridazine, **12**, and pyrazine, **13**. In these small molecules, spin-rotation interactions contribute

pyrimidine
(11)

pyridazine
(12)

pyrazine
(13)

to the ^{13}C spin–lattice relaxation rates over the entire temperature range studied. The dipolar contributions were determined by measuring the NOEs; then the ^{13}C dipolar relaxation rates are

$$R_1^{dd} = \frac{\eta}{1.988} (R_1^{dd} + R_1^{other}) \qquad (6.6)$$

For **11** and **12**, the data allowed interpretations in terms of molecular shape, attractive dipolar forces, and self-association of these molecules. The motion of pyrazine, **13**, however, could not be analyzed.

For polar molecules in relatively polar solvents, molecular reorientation may be slow because of specific solute–solvent interactions. Since the motional narrowing assumption may thus be invalid, Eq. 6.1 is not applicable, and more complex analyses must be used in these cases.[16,18]

If the motional behavior of a molecule can be predicted, its ^{13}C spin–lattice relaxation times may be useful in spectral assignments.[12,19,21,62,63] For example, T_1 values have aided in the assignments of ^{13}C spectra of a large number of chlorinated biphenylols,[64] assuming preferred rotation about the longitudinal molecular axis.[63] Wehrli[19] has written an extensive summary, which includes several examples featuring aromatic organic compounds, of the use of carbon T_1 values in assignment.

Spin-lattice Relaxation Times and Hindered Rotation

The effects of internal reorientations on ^{13}C T_1 values are closely related to the effects of overall molecular rotational anisotropy on these values. For example, dimethyldiphenylmethane, **14**, exhibits a methyl carbon T_1 value of 0.9 sec, in

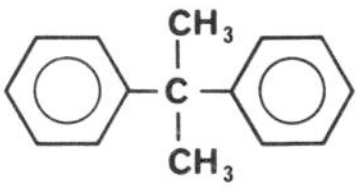

dimethyldiphenylmethane
(14)

marked contrast to toluene, which exhibits a T_1 of 16.3 sec for its freely rotating methyl carbon. The short T_1 in **14** is indicative of slow methyl rotation, since other proton-substituted carbons in this molecule have longer T_1 values, about 2–3 sec.

Several investigators have determined methyl group rotational barriers from carbon T_1 values, including calculations of such barriers for methyls in *o*-xylene, **15** (5.8 kJ/mol), for methyls at positions 1 and 3 in hemimellitene, **16** (6.06 kJ/mol), and for methyls at positions 1 and 3 in isodurene, **17** (6.48 kJ/mol).[47]

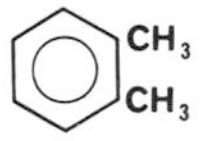

o-xylene   hemimellitene   isodurene
(15)   **(16)**   **(17)**

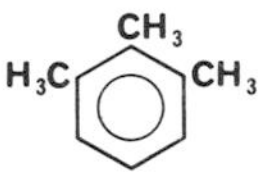

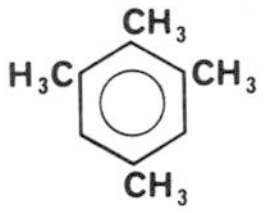

Note that the sixfold barriers for rotation of the 2-methyls in **16** and **17** will be characteristically low, so that these methyl groups rotate essentially freely.

Spin-lattice relaxation measurements in 1-methylnaphthalene, **18**, and 9-methylanthracene, **19**, give similar results.[21] For **18**, the methyl T_1 is roughly

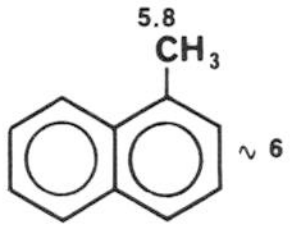

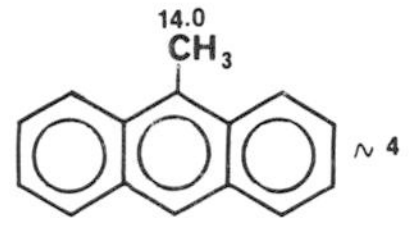

1-methylnaphthalene   9-methylanthracene
(18)   **(19)**

the same as that for the ring carbons, about 6 sec, as a result of the methyl steric interaction with the peri proton, H-8. In contrast, the methyl T_1 in **19** is much longer, reflecting its lower rotational barrier resulting from ground-state compression from steric interactions with two peri protons.[12]

Relative internal rotation of the two rings in some chlorinated biphenyl isomers (PCBs) is manifest in their T_1 values.[63] Some of these values are given in Figure 6.5. As in earlier studies of biphenyls, preferred rotation about the long molecular axis was observed for the PCBs. However, additional motional anisotropy is evident. For PCBs having one ring unsubstituted and chlorines at the ortho or meta positions in the substituted ring, shown at the left of the figure, the T_1 values are much longer for off-axis than for on-axis carbons in the unsubstituted ring. The longer values result from the faster rotation of this ring about the long axis, relative to the substituted ring. Chlorinated biphenyls having large hindrance to rotation about the inter-ring bond because of ortho–ortho' steric interactions, such as the two PCBs at the lower right of the figure, move nearly isotropically.

An interesting case of relative internal motions was observed in several substituted ferrocenes, **20**.[65] In the unsubstituted parent compound, the two rings

ferrocene

(20)

spin independently of each other and of the overall isotropic molecular tumbling. A substituent, R, on ring A slows the spinning of that ring relative to ring B. If ring B were to spin infinitely faster than ring A, the spin–lattice relaxation times of protonated carbons in ring B would be about four times those of similar car-

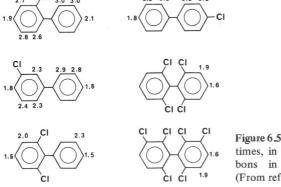

Figure 6.5. The ^{13}C spin–lattice relaxation times, in seconds, of the protonated carbons in some chlorinated biphenyls. (From reference 63.)

bons in ring A. The observed T_1 ratios (T_1^B/T_1^A) are 2 for R = acetyl and 2.4 for R = *n*-butyl, corresponding to spinning ratios for the unsubstituted to the substituted ring of 4 for acetylferrocene and 7 for *n*-butylferrocene.

Other Applications of Spin–Lattice Relaxation Times

In the spin–saturation method for the analysis of exchanging systems A $\leftrightharpoons$ B by ^{1}H NMR, introduced by Forsén and Hoffman some time ago,[66-69] the exchange rate and spin–lattice relaxation time of nuclei at site A are extractable from the time dependence of the NMR signal of nuclei at site B under conditions of A saturation. Unfortunately, the experiment is diffucult to perform and interpret with proton spectra, where extensive spin–spin coupling, nuclear Overhauser effects, and fast relaxation can create complexities in the data analysis and cloud its meaning. With ^{13}C spectroscopy, however, the problems are less severe, and result mainly from the low NMR sensitivity of ^{13}C.

For two-site exchange, in the Forsén–Hoffman spin–saturation treatment,

$$k_A = \frac{S_A^0 - S_A^\infty}{S_A^\infty T_{1A}} \tag{6.7}$$

where k_A is the rate of leaving site A; S_A^0 is the intensity of the A signal with no irradiation of B; S_A^∞ is the equilibrium intensity of the A signal a long time, $5(k_A + T_{1A}^{-1})^{-1}$, after saturation of A; and T_{1A} is the T_1 of nuclei at site A. If $T_{1A} = T_{1B}$, then standard inversion recovery T_1 experiments can be used to obtain the data. The spin–saturation method has several advantages: it makes accurate measurements of rates comparable to $1/T_1$, about 10^{-2} to 10^1 sec^{-1}, accessible; it does not require equally populated sites; and the spectra demonstrate clearly which sites are exchanging, since the strong irradiation of the A spin transitions obviously affects the B signal intensities by saturation transfer. So far, the method has not been applied extensively to aromatic systems. It does have significant potential, however, and one can expect such applications in the future. Unfortunately, a limiting feature of the spin–saturation method is the requirement of equal relaxation times at both sites.

A recent paper by Lambert and Keepers[70] details an approach to analysis of slow rate processes by ^{13}C spin–lattice relaxation time measurements in which equality of the T_1 values at the two sites is not required. This method is based on the coalescence of T_1 values. It supplies the exchange rate, k, and the true relaxation times for each site through regression analysis and curve fitting of the relaxation data to the double-exponential solution of the Bloch equations. By applying the method to dimethylformamide, the authors show that equivalent results are obtained for the activation parameters from this method and from earlier line shape analyses. Note that the rates measured are far smaller than those that give rise to line shape changes.

Various methods for the study of slowly exchanging systems, including extensions of the Forsén-Hoffman cross-saturation method and spin–echo-based T_2 methods, have been evaluated by Campbell and his co-workers.[71] For the model system chosen—the equilibrium between the two equivalent boat conformations of the seven-membered benzodiazepine ring in Valium®, **21**— longitudinal (T_1) relaxation measurements are shown to be superior to transverse (T_2) relaxation measurements.

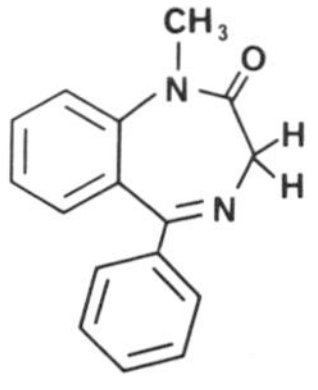

Valium
(21)

In another interesting application of ^{13}C T_1 measurements,[72] the predominant tautomer of 1-methylisoguanosine, **22**, in solution was established as the 2-keto, 6-amino form, **22a**. The authors assumed that the molecule was a rigid rotor and assumed a geometry, from which they calculated a value for the molecular rotational correlation time, τ_c. Since each tautomer—**22a**, **b**, or **c**—

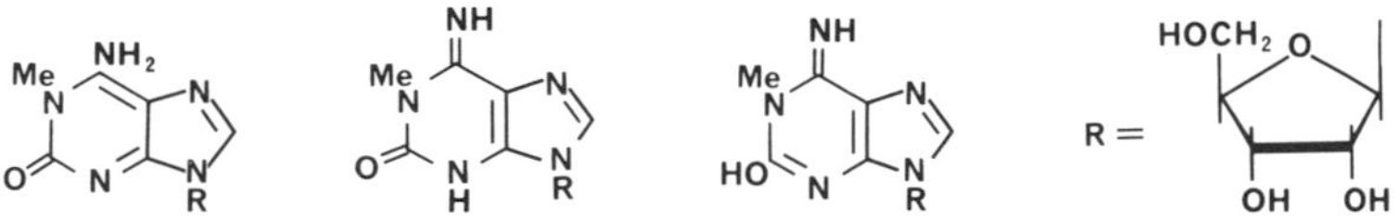

(22a) (22b) (22c)
1-methylisoguanosine
(22)

is unique with respect to the number of hydrogens two bonds removed from its quaternary carbons, the expected T_1^{dd} values differ for the three tautomers (because of the r^6 dependence of dipolar relaxation.) From the assumed internuclear distances and the calculated τ_c, the T_1^{dd} values were calculated and compared with experiment. Only tautomer **22a** had calculated quaternary carbon T_1^{dd} values that agreed with those measured.

Spin–Spin Relaxation Times and Chemical Exchange

Although direct spin–spin relaxation time measurements can, in principle, lead directly to exchange rates, in practice these measurements have been difficult to perform and interpret. The additional difficulties occurring in coupled abundant spin systems have prevented extensive T_2 applications in high-resolution proton spectroscopy. Applications in the past have employed primarily spin-echo techniques,[5] for which the theory and practice are well developed.[20,71,73]

The advent of pulsed Fourier transform NMR methods for routine observation of ^{13}C and other less abundant nuclei has brought about renewed interest in T_2 measurements on dynamic systems—in the form of $T_{1\rho}$ or "spin-locking" experiments. As we mentioned earlier in this chapter, for most medium-sized molecules in nonviscous solutions, $T_{1\rho} = T_2$.

Doddrell and his colleagues[74] have recently discussed the experimental requirements for, and the utility of, ^{13}C $T_{1\rho}$ measurements on dynamic processes in solution. The contribution to $T_{1\rho}$ from exchange for two equally populated exchanging sites, $T_{1\rho}^{ex}$, is given by

$$\frac{1}{T_{1\rho}^{ex}} = \frac{1}{4}(\Delta\omega)^2 \left[\frac{\tau_{ex}}{1 + \omega_1^2 \tau_{ex}^2}\right] \tag{6.8}$$

where $\Delta\omega$ is the chemical shift separation of the two sites in rad/sec, ω_1 is the strength of the spin–locking field, and $1/\tau_{ex}$ is the exchange rate constant. If chemical exchange is the only contributor to $T_{1\rho}$, the exchange contribution can be found from the relationship

$$\frac{1}{T_{1\rho}^{ex}} = \frac{1}{T_{1\rho}} - \frac{1}{T_1} \tag{6.9}$$

The rates that can be studied reliably by $T_{1\rho}$ methods are much faster than those accessible by spin–echo or line width experiments. One caveat should be kept in mind, however: broad-band proton heteronuclear decoupling, commonly used in ^{13}C NMR investigations, can cause rapid relaxation by scalar coupling of the ^{13}C spins in a spin–locking experiment. Thus it is essential that the decoupler be gated off during the spin–locking pulse.

As a demonstration of the power of these rotating frame experiments to provide useful results, the authors report determinations of the barrier to rotation of the CHO group about the partial double bond in benzaldehyde, **23**, and *p*-methoxybenzaldehyde, **24**. The $\Delta H^{\neq}$ values of 30.55 and 36.25 kJ/mol for

benzaldehyde   *p*-methoxybenzaldehyde
(23)   (24)

23 and **24**, respectively, were in good agreement with previous determinations in the literature. Significantly, $T_{1\rho}$ measurements allowed examination of the exchange process more than 100°K above the coalescence temperature.

CHEMICAL SHIFTS AND
LINE SHAPES OF SYSTEMS
INVOLVED IN DYNAMIC PROCESSES

Dynamic molecular processes whose effects are evident in NMR spectra can generally be divided into three categories—equilibria in which the rates of interconversion of the equilibrating species are slow on the NMR time scale, processes occurring at moderate rates such that the spectra of the interconverting forms are partially or completely averaged, and rapid processes manifested primarily in the relaxation times. In the preceding sections of this chapter, we discussed spin-lattice and spin-spin relaxation and presented some examples of their relevance to dynamic occurrences on the molecular level. Measurements of relaxation times, T_1 and T_2, have their greatest efficacy in the assessment either of fast events, evident mainly in T_1 values, or very slow events, evident primarily in T_2 values. Thus relaxation effects cover the ends of the rate range observable by NMR; the middle ground, encompassing those processes in the rate range of about 10 to 10^6 sec^{-1}, is covered by chemical shift and line shape changes. We now consider the principles of dynamic effects on NMR line shapes and chemical shifts. We then illustrate these effects with some results from recent publications.

The Effects of Exchange on Chemical Shifts and Line Shapes

In the slow exchange limit, the NMR spectrum of a sample in which there are several equilibrating species consists of a superposition of the spectra attributable to each individual species. The lifetime of a nucleus in a given magnetic environment or specific site is long enough, under slow exchange conditions, to allow several precessions of the nucleus at the frequency characteristic of that site before it jumps to another site. The nuclei at a given site produce resonance signals unique to that magnetic environment; nuclei at alternative sites likewise produce resonances distinctive of those environments. Since NMR absorption intensities, in the absence of complications from relaxation phenomena, are directly proportional to the numbers of nuclei giving rise to the signals, the populations of the various species present can be determined from the integrated signal intensities. Hence equilibrium constants and free energies can be determined from spectra at a single temperature. With slow-exchange spectra obtained at several temperatures, the activation parameters can be determined as well.

At the other end of the exchange rate continuum, when the interconversion of contributing species is fast, the observed NMR shielding of a particular set of nuclei is the average of their shieldings in the various environments, weighted according to the relative populations of nuclei in these environments. This averaging of the shieldings occurs as a result of the rapid transfer of the nuclei between sites, at a rate much greater than the difference in resonance frequencies of the sites, so that a given nucleus experiences the average of the different magnetic environments during the period of one precession. The observed chemical shift in the case of rapid exchange in thus

$$\delta = \sum_i n_i \delta_i \tag{6.10}$$

where δ_i is the chemical shift of nuclei at site i, and n_i is its fractional population.

Nuclear magnetic resonance chemical shifts normally exhibit a small temperature dependence, which can mostly be accounted for by temperature-dependent changes in the medium, such as density changes of the solvent. However, if a molecule of interest is involved in an exchange or conformational equilibrium, its chemical shift temperature dependence is much more severe and is often the first indication that some sort of averaging is taking place.

To extract the relative populations and hence obtain equilibrium constants from the averaged chemical shift values, one must have accurate knowledge of the chemical shifts of the individual species—not always an easy accomplishment. Model compound data have been used to predict individual chemical shifts, but this method is fraught with hazard.[75] A better approach is to examine the equilibrium at very low temperatures, below the slow-exchange limit, so that the signals from the contributing forms can be distinguished. The temperature dependences of the individual chemical shifts in the absence of significant exchange can then be extrapolated into the exchange region.

At intermediate rates, the effects of chemical exchange on NMR spectra are more complex. In simple cases, it may be possible to deduce exchange rates from line width measurements or chemical shift coalescences, as described below. Fairly often, however, one must resort to full line shape analysis, which has been treated in detail elsewhere.[24,25,76]

For exchange of uncoupled nuclei between two magnetic environments characterized by resonance frequencies ν_A and ν_B Hz, equal populations, equal values of T_2, lifetimes in those environments τ_A and τ_B, and a mean lifetime for exchange defined as $\tau = \tau_A \tau_B / (\tau_A + \tau_B)$, the NMR spectrum changes with the exchange rate as follows. When $\tau \gg 1/(\nu_B - \nu_A)$, the individual resonances at ν_A and ν_B are observed. As the exchange lifetime decreases, these signals, which initially had line widths, $\Delta\nu$, determined by T_2^*, broaden and gradually move toward each other until they coalesce to a single broad peak. The mean lifetime at the point when coalescence occurs is

$$\tau = \frac{\sqrt{2}}{2\pi(\nu_B - \nu_A)} \tag{6.11}$$

Further decreases in τ bring about narrowing of the coalesced peak, until the limiting line width, determined by T_2^*, is reached. At this limiting line width, the condition for fast exchange, $\tau \ll 1/(\nu_B - \nu_A)$, is fulfilled.

In the rate ranges above or below coalescence, but not at the fast- or slow-exchange limits, rates may be extracted from the line widths. Below coalescence, where the peaks are broadened but do not yet overlap,

$$\frac{1}{\tau} = 2\pi(\Delta\nu' - \Delta\nu) \tag{6.12}$$

where $\Delta\nu'$ is the width of the broadened line. Since the lifetime of nuclei at a single site is involved, Eq. 6.12 may also be used for many-site exchange, provided there is no overlap of the resonances. Above coalescence, rates may be obtained from the line widths using the relationship

$$\frac{1}{\tau} = \frac{\pi(\nu_B - \nu_A)^2}{(\Delta\nu' - \Delta\nu)} \tag{6.13}$$

Both of the relationships between exchange rates $1/\tau$ and line widths given in Eqs. 6.12 and 6.13 are approximate, and are invalid when the exchange rate is of the order of the chemical shift difference between sites—that is, when τ is approximately $\sqrt{2}/2\pi(\nu_B - \nu_A)$. Near coalescence, the lifetimes must be determined by total line shape analysis, using the complete equations governing the line shape.[24,25,76]

Applications of Dynamic NMR to Aromatic Organic Compounds

Tautomerism

The NMR spectrum at the slow-exchange limit can often establish the presence of several equilibrating forms or one energetically favored form of a substance. Tautomeric equilibria frequently lie in the slow-exchange region and thus are amenable to analysis by the techniques discussed in the preceding section of this chapter. An interesting example of tautomerism is that in a series of 1-phenyl-pyrazolin-5-ones, **25**.[77] In the ^{1}H spectrum, only one tautomer was apparent. For R = H, CH$_3$, and COOC$_2$H$_5$, the ^{1}H spectrum ruled out structure **25a**, but did not allow distinction between **25b** and **25c**, since proton NH and OH resonances are

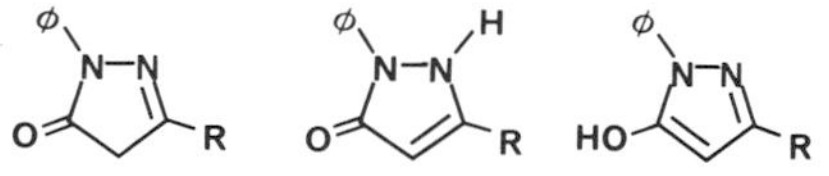

(25a)   (25b)   (25c)

quite similar and proton exchange between these groups could occur. The ^{13}C spectrum clearly established these tautomers as **25c**, by the absence of both a ring CH$_2$ and a carbonyl resonance. In contrast, the R = NH$_2$ tautomer exists as

25a, demonstrated by the presence of both CH_2 and $C = O$ resonances in its ^{13}C spectrum.

A study of 17 indazole derivatives substituted in various positions—**26, 27, 28,** and **29**—established the effects of the different substitution patterns on the

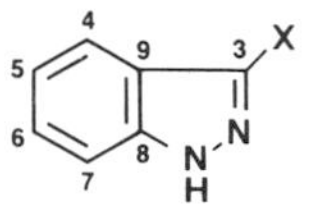

3-X-indazoles
(26)

^{13}C chemical shifts of these compounds.[78] By comparison of the chemical shifts of N–H indazoles, **27,** with those of 1-methyl- and 2-methylindazoles, **28** and **29,**

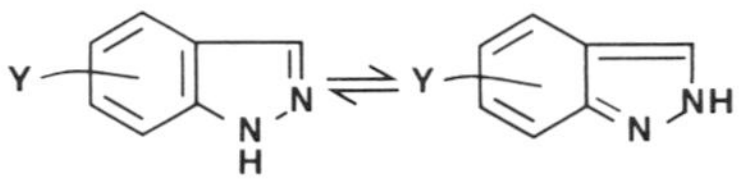

N–H indazoles   1-methylindazoles
(27a)   **(27b)**   **(28)**

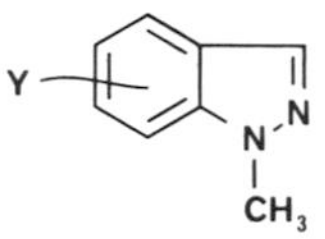

the predominant structures of **27** are clearly established as the N–1–H tautomers **27a**. The chemical shift for C-8 in **27,** for example, ranges from δ_C 139.9 to δ_C 141.6, in comparison with δ_C 137.9 to 141.3 for this carbon in **28,** and in contrast with δ_C 147.8 to 149.1 for this carbon in **29** (except for the 7-NO_2 compound, in which the α nitro group increases the C-8 shielding to δ_C 139.6).

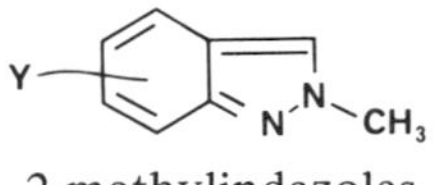

2-methylindazoles
(29)

Several derivatives of N-hydroxybenzotriazole, **30,** were examined by ^{13}C NMR to evaluate the tautomeric equilibrium of **30** and the isomerism of its

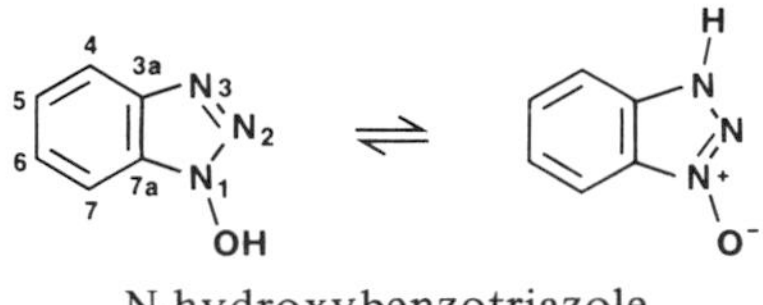

N-hydroxybenzotriazole
(30a)   **(30b)**

N-acylated derivatives.[79] Compound **30** is known from extensive UV spectroscopic studies[80] to be about 80% **30a** in ethanol, whereas the N-oxide form,

30b, predominates in water. The major tautomer is strongly dependent on the nature of the solvent, with **30a** favored in organic media. By comparisons of the ^{13}C chemical shifts of **30** with those of its derivatives, **31–35**, the benzoyl derivative was shown to have the O-acyl structure (**33a**), the methyl carbonate ester

N-hydroxybenzotriazole derivatives

(**31**)	R = CH_3	(**32**)	R = CH_3
(**33a**)	R = C_6H_5CO	(**33b**)	R = C_6H_5CO
(**34a**)	R = CH_3OCO	(**34b**)	R = CH_3OCO
(**35a**)	R = C_6H_5OCO	(**35b**)	R = C_6H_5OCO

to exist mostly as the N-acyl isomer (**34b**), and the phenyl carbonate ester to exist as a mixture of the two isomers **35a** and **35b**. From its greater chemical shift similarity to **31** than to **32**, the parent compound, **30**, was clearly shown to be the N-hydroxy tautomer (**30a**). The chemical shifts of the latter three compounds are given in Table 6.1.

A similar study of *s*-triazolo-*as*-triazinones[81] employed both ^{13}C chemical shifts and ^{13}C–^{1}H coupling constants to investigate the structures of five isomeric series of these compounds. Both the type of ring junction between the two heterocycles and the predominant tautomeric form in each isomeric system could be established.

^{13}C NMR does not always allow observation and interpretation of tautomeric equilibria, however. Elguero and his co-workers[82] carried out a systematic study of the effects of substitution, lanthanide shift reagents, and solvent changes on the ^{13}C chemical shifts of pyrrole, pyrazole, imidazole, *s*- and *v*-triazole, and

Table 6.1 ^{13}C **Chemical Shifts of Some N-Hydroxybenzotriazole Derivatives in Dimethyl-d_6 Sulfoxide**

Compound	δ_C						
	C-4	C-5	C-6	C-7	C-3a	C-7a	CH_3
N-Hydroxybenzo-triazole, **30**	119.4	124.6	127.3	109.8	142.9	128.1	
1-Methoxybenzo-triazole, **31**	119.7	125.0	128.4	109.2	143.0	126.4	68.1
3-Methylbenzo-triazole-1-oxide, **32**	111.7	130.1	124.6	114.6	134.3	129.3	34.5

Source. Reference 79.

tetrazole. Their aim was to assess the utility of ^{13}C NMR in the determination of the positions of tautomeric equilibria for rapidly interconverting tautomers, particularly as compared to ^1H or ^{14}N spectroscopy. Unfortunately, these ambient temperature studies failed to illumine the nature of the tautomerism, even with model compound data available. The authors concluded that ^{13}C NMR offered no significant improvement over ^1H or ^{14}N NMR in the resolution of the equilibria of rapidly interconverting azole tautomers.

A later investigation of tautomerism in pyrazoles presented an answer to the problem of extremely fast proton exchange in azoles. Using hexamethyl phosphoramide as a solvent, Chenon and her colleagues[83] were able to reduce the rate of tautomeric exchange in pyrazole, **36,** so that separate signals were observed

pyrazole tautomers
(36)

in the ^{13}C spectrum for C-3 and C-5 at ambient temperature. At 47°C, these signals coalesced, giving a $\Delta G^{\ddagger}$ of 15 kcal/mol from the corresponding exchange rate (see Eq. 6.11). The ^1H spectrum did not exhibit separate lines for the tautomers, even at −17°C in hexamethyl phosphoramide.

With decreasingly basic solvents (dimethyl sulfoxide, acetone) the exchange rates were greater at a given temperature both for **36** and for the substituted pyrazoles in the study. The trend of increased activation energy for tautomeric exchange with increased solvent basicity was attributed to solvent-stabilized tautomeric structures in the more basic solvents. Although hexamethyl phosphoramide is a somewhat hazardous solvent, its effectiveness in shifting tautomeric equilibria into a convenient and easily accessible rate range warrants its use in NMR studies of rapid exchange in tautomeric compounds.

Deuterium isotope effects on the ^{13}C chemical shifts of some enamino ketones have been used recently[84] to paint a qualitative picture of tautomeric equilibria in these compounds. The enamino ketones exist as one or more of the potential tautomers 37. In the experiment, spectra of the compound of interest

enamino ketone tautomers

(37a)   (37b)   (37c)

in both D_2O and in H_2O are obtained at the same time, employing coaxial sample tubes.[85] The isotope-shifted resonances are thus superimposed on the spectrum of the nondeuterated compound, as is illustrated in Figure 6.6.

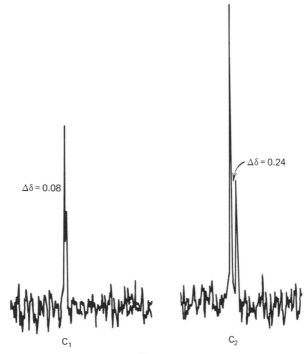

Figure 6.6. The low-field region of the ^{13}C NMR spectrum of the enamino ketone **39**, showing the deuterium isotope-shifted resonances of carbons 1 and 2. (From reference 84.)

Deuterium isotope effects on the ^{13}C shieldings of geminal carbons, through two bonds, generally increase the shieldings of these carbons by about 0.1 to 0.2 ppm. On vicinal carbons, these isotope shifts are usually small or negligible. Therefore, if the predominant tautomers of the enamino ketones **38** to **40** are

      (38)   (39)   (40)

of type **37a**, isotope shifts toward higher field will be observed for C-2 and C-α; if of type **37b**, such a shift will be observed for C-1; and if of type **37c**, such shifts will be observed for C-1 and C-2, in addition to splitting of the ipso carbon (C-4 in **37c**) resonance with $J_{CD} \geqslant 18$ Hz and an ipso isotope shift of 0.3–0.7

ppm. Structure **37c** can also be distinguished from the other two tautomers, **37a** and **37b**, by the sp^3 rather than sp^2 resonance of C-4.

The low-field region of the ^{13}C spectrum of **39**, shown in Figure 6.6, clearly establishes the presence of both tautomeric forms **37a** and **37b**, since isotope shifts are evident for both C-1 and C-2. A similar result was obtained for **40**. For compound **38**, however, deuterium isotope effects were observed only at C-2, C-α, and C-4, suggesting that only form **37a** is present. The latter result is in agreement with the known preference of the double bond to be exo to a five-membered ring.

Several other NMR studies of tautomerism in aromatic heterocycles are in the scientific literature. These include a structural determination of 3-azidoindazole by ^{1}H and ^{13}C NMR,[86] determinations of the tautomeric populations of some purines by ^{13}C NMR methods,[87,88] and studies of tautomerism in some amino-indenes.[89]

Alei and his co-workers[90] have extended several earlier studies[91,92] in an investigation of the effects of the pH-dependent protonation of imidazole, **41**, and 1-methylimidazole, **42**, on their ^{15}N NMR parameters in aqueous solution.

imidazole

(41)

1-methylimidazole

(42)

In the pH range 5-9, where significant amounts of both imidazole and imida-zolium ion are present, rapid interconversion of these species results in a single ^{15}N resonance for **41**, with the frequency and line width dependent on pH. From the line widths, the proton exchange rate was deduced to be k $\simeq 10^4$ sec$^{-1} M^{-1}$. At pH greater than 10 or less than 4, only one **41** species is present, either the neutral or the protonated form, respectively. Rapid exchange of ^{15}N–H protons with water precluded observation of directly bonded ^{15}N–^{1}H coupling, but well-resolved multiplet structures of the averaged ^{15}N resonances caused by coupling with the C–H ring protons were observed in both the ion and the neutral molecule. At very high acidities, $>6M$ HCl, proton exchange with water was sufficiently slowed that the directly bonded ^{15}N–^{1}H coupling could be seen. Spectra of 1-methylimidazole, **42**, illustrating this behavior are shown in Figure 6.7. The authors also report ^{13}C–^{15}N, ^{15}N–^{15}N, and ^{1}H–^{1}H coupling constants in these compounds, and discuss the implications of this study for NMR studies of histidines.

Related studies of protonation, proton exchange, and pH dependences of aromatic hydrocarbons and heterocycles are numerous.[3,4,5,29,93] These studies include measurements of the pH dependences of the ^{13}C spectra of several

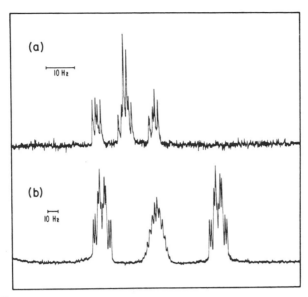

Figure 6.7. ^{15}N NMR spectra at 10.16 MHz for aqueous 1-methylimidazole, **42**. (*a*) N–3 spectrum of 2 *M* neutral **42** at pH 12.7. (*b*) N–1 and N–3 (doublet) spectrum of 1.7 *M* protonated **42** in 8.5 *M* HCL. (From reference 88.)

pteridines, including folic acid, for which the pK_a value was determined;[94] ^{13}C NMR investigations of the sites of protonation and pK_a values of some diazanaphthalenes;[95] and related ^{13}C studies of protonation of several nitrogen heterocycles.[96]

Stable Carbocations and Carbanions of Aromatic Hydrocarbons

Arenium ions, which are well established as intermediates in electrophilic substitution reactions and in many acid-catalyzed transformations of aromatic compounds, have been studied extensively over the years to elucidate their electronic structures. Until recently, however, the simplest of these—the benzenium ion, **43**—has eluded characterization as a static (nonequilibrating) species by NMR spectroscopy.[97] Full characterization of the naphthalenium, **44**, and anthracenium, **45**, ions has also been lacking.[98,99]

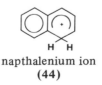
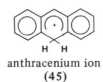

benzenium ion   napthalenium ion   anthracenium ion
  (43)   (44)   (45)

Olah and his colleagues, in an extensive series of papers, have reported NMR spectroscopic observations of numerous carbocations, stabilized in superacid solution and often at low temperatures. ^{13}C NMR is generally the most suitable technique for the structural study and determination of the charge distributions in carbocations. Thus the Olah group[100] employed ^{1}H and ^{13}C nuclear magnetic resonance to study the arenium ions (**43, 44,** and **45**).

The initial studies of the benzenium ion[97] under superacidic non-nucleophilic conditions gave temperature-dependent averaged 100 MHz ^{1}H and 25.1 MHz ^{13}C spectra, which corresponded to a set of benzenium ions, **46a–f**, rapidly interconverting, even at −135°C, by a series of 1,2 shifts. Therefore, later studies

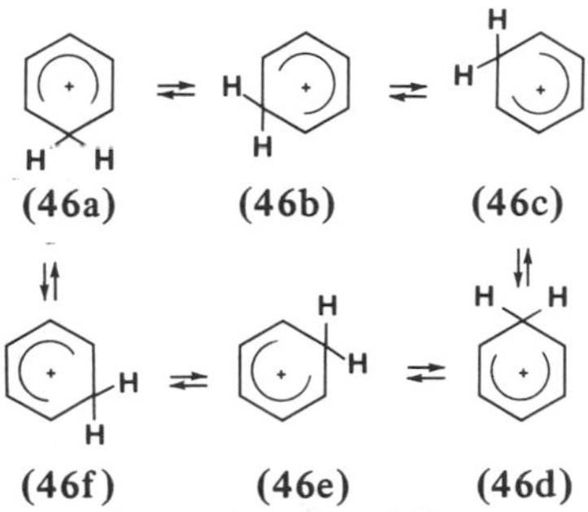

(46a)   (46b)   (46c)

(46f)   (46e)   (46d)

benzenium ion shifts

were done at higher magnetic fields, since the exchange rate can be related directly to the frequency difference between exchanging sites, as indicated in Eq. 6.11. With the higher applied fields, 270 MHz for ^{1}H and 67.89 MHz for ^{13}C, and at −140°C, spectra of the static benzenium ion, **43**, were observed. The ^{1}H NMR spectrum of **43** at high field and low temperature is shown in Figure 6.8. Even under these conditions, because of limited equilibration of **46a–f**, fine structure of the resonances is not apparent. The three broad ^{1}H resonances in the figure correspond to overlapping peaks for H-1, -5, and H-3 at δ_H 9.7; a peak for H-2, -4 at δ_H 8.6; and a peak for H-6, -6' at δ_H 5.6.

Comparisons of the static spectra with partially coalesced spectra obtained earlier[97] and the approximate C_{2v} symmetry of **43** support an Arrhenius activation energy, E_a = 10 kcal/mol, for the equilibration of **46a–f**. This E_a value approximates the energy difference between the parent benzenium ion, **43**, and the benzonium ion transition state, **47**, for the equilibration process.

benzonium ion
(47)

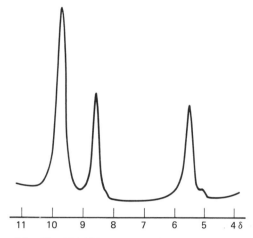

Figure 6.8. The 270 MHz proton NMR spectrum of the non-equilibrating benzenium ion, **43**, in $SbF_5-FSO_3H-SO_2ClF-SO_2F_2$ solution at $-140°C$. (From reference 100.)

The ^{13}C NMR chemical shifts of the static benzenium ion—C-1, -5, δ_C 186.6; C-2, -4, δ_C 136.9; C-3, δ_C 178.1; and C-6, δ_C 52.2—could be interpreted in terms of its positive charge distribution. They suggest approximate C_{2v} symmetry and the absence of antihomoaromatic character in this system.

Similar results for the naphthalenium, **44**, and anthracenium, **45**, ions permitted interpretation of the ^{13}C spectra of these species in terms of charge density distributions. Comparisons of the ^{13}C shieldings of **44** and **45** with those obtained for series of substituted naphthalenium and anthracenium ions allowed evaluation of the influences of substituents on the electronic structures of arenium ions.

The compound heptalene, **48**, is extremely unstable. Oth and his co-workers[101,102] were able to establish, by ^{13}C NMR spectroscopy at temperatures

heptalene
(48)

between $-100°$ and $-167°C$, that heptalene in its ground state has fixed π bonds but undergoes extremely rapid 1,2 shifts. The ^{13}C spectrum of **48** at $-167°C$ exhibits six signals, each corresponding to two equivalent carbons, and indicates fixed π bonds at very low temperatures, as is shown in the representative structure **49**, in which equivalent carbons are denoted by identical letters. In this state, heptalene most probably is nonplanar and chiral with C_2 symmetry as

heptalene
equivalent carbons
(49)

shown in Figure 6.9. At $-100°$, the fast bond shift process, **50**, averages the number of ^{13}C NMR signals of heptalene to four: C-1, -5, -6, and -10, δ_C 137.4; C-2, -4, -7, and -9, δ_C 132.8; C-3 and -8, δ_C 133.3; and C-5a and -10a, δ_C 143.1. If **50** is nonplanar, as in Figure 6.9, additional averaging may occur by isodynamical double ring inversion.

heptalene interconversion
(50)

The dianion of heptalene possesses 14 π electrons, and thus a planar aromatic structure of D_{2h} symmetry is highly likely for this species. Low-temperature ^{1}H and ^{13}C NMR spectroscopy[101] indicate such a structure. The proton NMR spectrum of the dianion is nearly first-order, with chemical shifts H-1, -5, -6, and -10, δ_H 7.41; H-2, -4, -7, and -9, δ_H 5.64; and H-3 and -8, δ_H 6.13, and coupling constants J_{12} = 9.25; J_{34} = 8.70; and J_{13} and J_{24} $<$ 0.2 Hz. Since the spectrum is only slightly temperature-dependent from $-80°$ to $+100°$C, it is unlikely that chemical shift averaging is occurring. Thus the dianion is aromatic, with structure **51**.

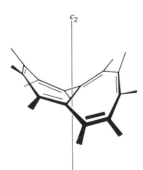

Figure 6.9. The most probable conformation of heptalene. (From reference 100.)

heptalene dianion
(**51**)

The ^{13}C chemical shifts of the dianion, **51**, are quite dissimilar to those of the neutral molecule, **48**. In **51**, only four signals are observed throughout the temperature range. The corresponding ^{13}C chemical shifts are C-1, -5, -6, and -10, δ_C 111.1; C-2, -4, -7, and -9, δ_C 91.1; C-3 and -8, δ_C 103.8; and C-5a and -10a, δ_C 113.9. Hence the ^{13}C spectrum supports the planar aromatic structure deduced for the dianion from the ^{1}H NMR results, with 14 π electrons delocalized over twelve p_z orbitals.

Several other studies of charge distributions and ^{13}C NMR parameters in ions of aromatic molecules have been reported, and are reviewed elsewhere.[13,28,31]

Complex Equilibria

Chemical shift methods for studies of complex equilibria in the fast-exchange region have been applied extensively.[5] Using the general relationship that the observed chemical shift is the population-weighted average of the chemical shifts of the contributing species expressed in Eq. 6.10, for a rapid equilibrium between two species, complexed and noncomplexed,

$$A + B \rightleftharpoons A \cdot B \qquad (6.14)$$

the observed chemical shift, δ, for a given set of A nuclei is

$$\delta = x_{AB}\delta_{AB} + x_A\delta_A \qquad (6.15)$$

where x_{AB} and x_A are the mole fractions and δ_{AB} and δ_A are the chemical shifts of A nuclei for the complex and noncomplexed A, respectively. If the concentration of complexing agent B is in great excess, the A chemical shifts and the equilibrium constant, K, are related by

$$\frac{\delta - \delta_A}{[B]} = -K(\delta - \delta_A) + K(\delta_{AB} - \delta_A) \qquad (6.16)$$

which gives a linear plot of $(\delta - \delta_A)/[B]$ vs. $(\delta - \delta_A)$ for 1:1 stoichiometry, with slope $-K$ and intercept $K(\delta_{AB} - \delta_A)$.[103] When higher order complexes are present, more complicated expressions for the dependence of δ on K and [B], which are amenable to nonlinear regression analysis, apply.[104] These methods are detailed in a proton NMR study of complex equilibria involving the pesticide DDT–1,1,1-trichloro-2,2-bis(*p*-chlorophenyl)ethane, **52**.[104,105] The study

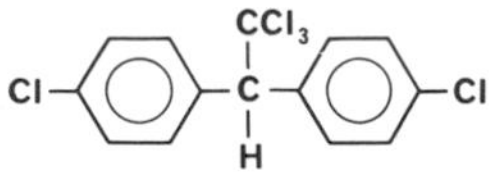

1,1,1-trichloro-2,2-bis(*p*-chlorophenyl)ethane
(52)

demonstrated the participation of **52** in π-complex formation with aromatic π-electron donors, such as benzene and naphthalene, and association of **52** with polar complexing agents, such as ethyl acetate, because of the highly polar benzhydryl C–H group of **52**. Thermodynamic parameters for several complex equilibria between **52** and various types of complexing agents were reported.

Donor–acceptor complex formation between methyltin trichloride and several 4-substituted pyridines has been similarly studied by ^{1}H NMR.[106] The formation of 1:1 and 1:2 MeSnCl$_3$:pyridine complexes was proved, and equilibrium constants were extracted from the data.

^{15}N chemical shift studies have been used to evaluate complex formation between Zn^{2+} and imidazole, **41**,[92] and between Cd^{2+} and imidazole[91] in aqueous solution. In the latter study, the number of imidazoles bound to the cadmium cation was determined from the average nitrogen chemical shifts and previously determined stepwise association constants. Coordination to Cd^{2+} produces a diamagnetic shift of 8–12 ppm in the ^{15}N resonance of **41** relative to neutral aqueous **41**. Evidence for coordination of no more than four imidazole molecules per Cd^{2+} was obtained,[91] in contrast to six imidazoles per Zn^{2+} in the prior study.[92]

The structures of the complexes of benzo[*a*]pyrene, **53**, and the free radical 6-oxybenzo[*a*]pyrene, **54**, with caffeine, **55**, have been studied by ^{1}H NMR,

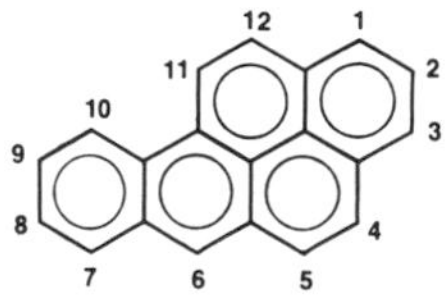

| benzo[*a*]pyrene | 6-oxybenzo[*a*]pyrene | caffeine |
| (53) | (54) | (55) |

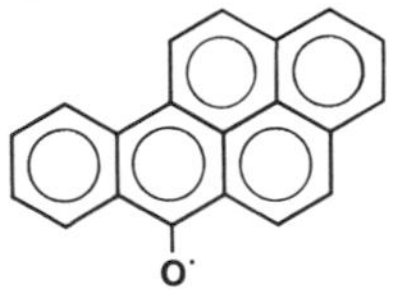

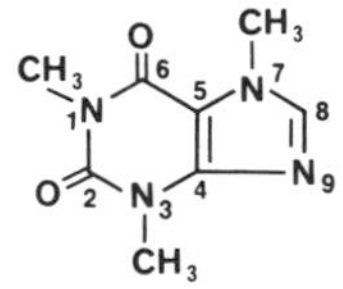

using both chemical shift and spin–lattice relaxation time measurements.[107] From ring-current effects on the **53** and **55** chemical shifts, and from the concentration dependences of these shifts, a 1:2 benzo[*a*]pyrene:caffeine sandwich-type complex was deduced. Support for this complex was gained from the proton spin–lattice relaxation times (T_1 values) of the benzo[*a*]pyrene in caffeine–D$_2$O solution. Assuming all proton T_1 values were the same, and assuming single-exponential relaxation, a rotational correlation time $\tau_c = 1.1 \times$

10^{-10} sec was calculated for the complex. This τ_c value was approximately 3-4 times that calculated for caffeine alone, $\tau_c = 3 \times 10^{-11}$ sec. Since τ_c values should be proportional to the molecular volumes, hence the molecular weights, the relaxation data lend credence to the 1:2 complex between **53** and **55**. Data for complex formation between the radical, **54**, and caffeine suggest structures analogous to those of the parent compound, **53**, and caffeine.

Charge-transfer complexes between the π-electron acceptor trinitrobenzene and the π-electron donors benzene, naphthalene, and anthracene have been studied by ^{13}C NMR,[108] as have similar complexes between o-chloranil and 1-methylnaphthalene.[109,110] Association constants and the shifts of the carbons in the 1:1 complexes were determined. In the acceptor molecules, complexation led to diamagnetic shifts (increased shielding) of the ^{13}C resonances. However, in the donors, both diamagnetic and paramagnetic shifts of the carbon signals were observed. These differences may prove useful in establishing the sites of interaction and perhaps other characteristics of charge-transfer pairs.[108,111]

Numerous investigations have concerned aromatic complexes with various metal ions and organometallic ligands. The effects of divalent paramagnetic metal acetylacetonates on the ^{13}C magnetic resonance parameters of pyridine, quinoline, and isoquinoline were investigated.[112] Specific interactions between the aromatic heterocycles and nickel, copper, cobalt, and manganese acetylacetonates were characterized. Another study sought to determine by ^{13}C NMR the hapto properties of several iron, chromium, nickel, and cobalt ligands in their π complexes with indene.[113] In a similar vein, ^{13}C spectra were used to characterize π-(arene) tricarbonylchromium complexes.[114]

Fluxional Molecules and Organometallic Rearrangements

Participation of certain organometallic compounds in rapid permutational isomerization reactions (fluxionality) can be monitored readily by NMR techniques.[12] One of the first of such substances to be studied by ^{1}H NMR was the fluxional molecule $(h^5\text{-}C_5H_5)(CO)_2Fe(h^1\text{-}C_5H_5)$, **56**. Degenerate rearrangements in

$$(h^5\text{-}C_5H_5)(CO)_2Fe(h^1\text{-}C_5H_5)$$
$$(56)$$

this system proceed by a series of 1,2 shifts of the iron-carbon σ bond as is shown in **57**. Variable-temperature ^{13}C spectra[115] also support this mechanism.

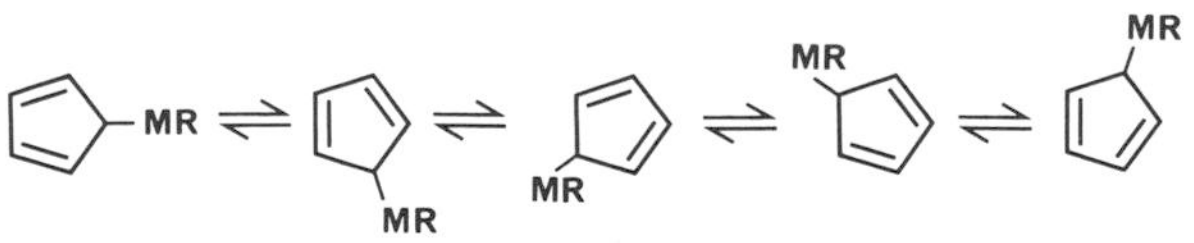

metal-carbon σ bond shifts
(57)

At ambient temperature, three resonances were observed—a single resonance for the π-bonded ring, a single resonance for the two carbonyls, and a broadened resonance for the σ-bonded ring. At dry ice temperature, $-78°C$, the broad signal is resolved into three peaks, with a $2:2:1$ intensity ratio. The C-1, -4 signal sharpens more slowly with decreasing temperature than does the C-2, -3 signal. From -78 to $-88°C$, the C-1, -4 resonance broadens again, because of hindered rotation about the Fe—C-5 bond leading to nonequivalence of C-1 and C-4 and of C-2 and C-3. The fluxionality can be described as a three-site exchange process, the sites being 1 and 4, 2 and 3, and 5, with $2:2:1$ population ratios. Calculated spectra for this process agreed well with experiment, and allowed extraction of the Arrhenius activation energy for the exchange, $E_a = 10.7 \pm 0.5$ kcal/mol, from line shape analysis.

The mechanism of the intramolecular metallotropic rearrangement of trimethylstannyl indene, **58**, has been determined[116] by ^{13}C NMR. The fluxional

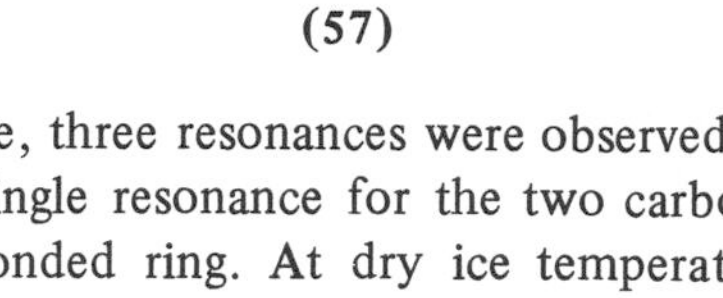

1-trimethylstannyl-1*H*-indene
(58)

behavior could proceed through either two successive 1,2 shifts or one 1,3 shift. For the former mechanism, 1,2 shifts would require the compound to go through the intermediate, **59**, which is about 9 kcal/mol less stable than **58**. This

2-trimethylstannyl-2*H*-indene
(59)

would make $\Delta G^{\ddagger}$ much higher for **58** than for the analogous cyclopentadienyl compound, **60**, if the 1,2 mechanism were operative in both. For the latter

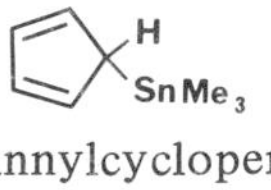

1-trimethylstannylcyclopenta-2,4-diene
(60)

mechanism, a 1,3 shift, $\Delta G^{\ddagger}$ values should be about the same for **58** and **60**. From the temperature dependence of the C-8, -9 and C-4, -7 resonance line widths, E_a = 13.8 ± 0.8 kcal/mol for **58**, whereas for **60**, E_a is 6.8 ± 0.7 kcal/mol.[109] The data thus establish the 1,2 shift mechanism.

Other examinations of fluxionality and organometallic rearrangements include one of organosilicon and organotin indenyl derivatives,[117] of Group IVA-substituted indenes and indenyl anions,[118] and of carbonyl scrambling in azulenepentacarbonyldiiron and its ruthenium analog.[119]

Steric Hindrance and Rotational Barriers

Barriers to rotation about single bonds in sterically hindered molecules are often of a magnitude such that their effects are visible on the line shapes in NMR spectra obtained in an experimentally accessible temperature range. An interesting example is the hindered rotation about the central C-9–C-9' bond in 9,9'-bifluorenyls, **61**, studied by Olah and his co-workers.[120] At low temperatures

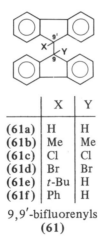

	X	Y
(61a)	H	H
(61b)	Me	Me
(61c)	Cl	Cl
(61d)	Br	Br
(61e)	*t*-Bu	H
(61f)	Ph	H

9,9'-bifluorenyls
(**61**)

and in the solid state, these compounds are frozen as the gauche conformers, one of which is shown as **62**. This preferred gauche conformation is general for members of the class of "clamped" polyarylethanes, in which the aryl rings are

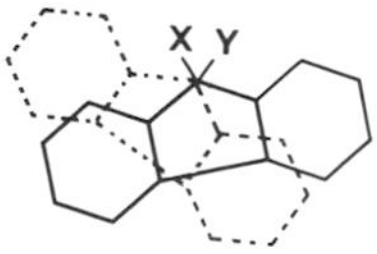

gauche conformer of
9,9'-bifluorenyl
(**62**)

constrained so that the ring planes are held away from the central sp^3-sp^3 bond. Since there are two gauche conformers, the low-temperature ^1H NMR spectra consist of superpositions of the two independent conformer sub-spectra. From line shape analysis of the broadening and coalescing spectra obtained as the temperature was increased, the Olah group[120] was able to extract exchange lifetimes and hence the rotational barriers for conformer interconversion. This interconversion of the gauche rotamers may go through either an anti conformation, **61**, or an eclipsed form, **63**. The measured barriers are given in Table 6.2. For the

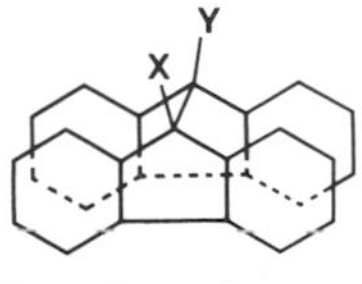

eclipsed conformer of
9,9'-bifluorenyl
(63)

methyl derivative **61b**, $\Delta G^{\ddagger}$ is smaller than expected, but the origin of the barrier may be obscured because it is the *difference* in energy between the ground and transition states that is measured. In any event, the preferred chiral gauche ground state conformers, with hindered rotation about C-9–C-9', are supported by the NMR data. Evidence was also obtained of restricted rotation about the axis of the 9-*t*-butyl group in **61e**, with $\Delta G^{\ddagger}$ = 14.4 kcal/mol.

A related study[121] employed ^{13}C NMR line shape analysis to examine the sterically restricted rotation of the ethyl group in some 9-ethyltriptycenes, **64–66**. Because the α carbon substituents are not identical, these compounds

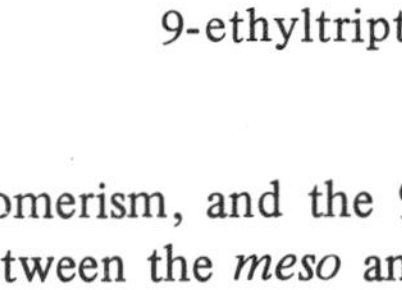

	W	X	Z
(64)	H	OMe	OMe
(65)	Cl	Cl	H
(66)	H	Me	Me

9-ethyltriptycene derivatives

exhibit optical isomerism, and the 9-ethyl group rotation is equivalent to an interconversion between the *meso* and *dl* isomers, as shown in the Newman projections, **67**. This interconversion is relatively slow on the NMR time scale,

Table 6.2. Barriers to Rotation
About the C-9–C-9′ Bond in
9,9′-Bifluorenyls

Compound	$\Delta G_{240}^{\ddagger}$, kcal/mol
61a	9.9
61b	10.0
61c	13.2
61d	13.2
61e	11.2

Source. Reference 120.

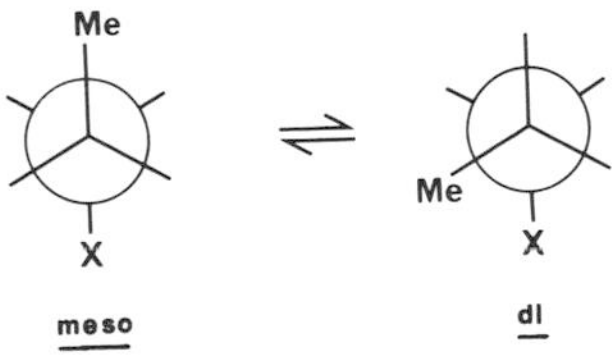

Newman projections of
9-ethyltriptycene
(67)

owing to the strong repulsion between the ethyl group and peri substituents on the rigid triptycene skeleton, and at moderately low temperatures, both forms are clearly evident in the spectra. For example, the spectra of the aromatic carbons 2 and 3 in **64** shown in Figure 6.10 exhibit separate sharp peaks for the *meso* and *dl* isomers. At $-34.3°C$, the signals a and d from the *meso* isomer are in the intensity ratio 60:40 to the signals b and c from the *dl* isomer. This result agrees with the greater stability expected for the *meso* isomer because of its lesser steric repulsion between the ethyl and methoxy groups.

From computer simulations of the ^{13}C line shapes of **64**, the relative populations of the conformers and the static thermodynamic parameters for ethyl group rotations were obtained. These are $\Delta G^0 = 0.12 \pm 0.13$ kcal/mol, $\Delta H^0 = 0.80 \pm 0.13$ kcal/mol, and $\Delta S^0 = 2.3 \pm 0.4$ eu. The rotational barrier, $\Delta G^{\ddagger} = 13.8 \pm 0.3$ kcal/mol, for interconversion of the *meso* and *dl* forms was likewise obtained. It compares with $\Delta G^{\ddagger}$ values of 8.8 kcal/mol for 9-methyl-, 25.4 kcal/mol for 9-*i*-propyl $[d(l) \rightarrow l(d)]$, and >30 kcal/mol for *t*-butyl-triptycenes.

Rotational barriers about the C_{Ar}–C_9 bond in several 9-aryltriptycenes, including the *o*-tolyl and *o*-anisyl derivatives, **68**, have also been studied.[122]

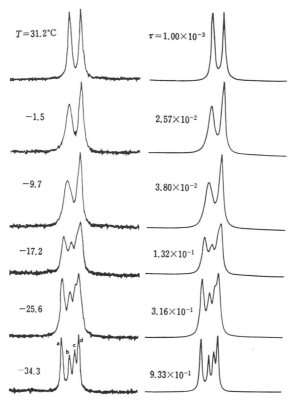

Figure 6.10. The observed (left) and calculated (right) ^{13}C spectra of the aromatic carbons 2 and 3 in 1,4-dimethoxy-8-ethyltriptycene, **64**, at several temperatures. (From reference 121.)

9-*o*-tolyl- and 9-*o*-anisyltriptycene
(68)

From the temperature dependences of the proton NMR spectra, rotational barriers $\Delta G^{\ddagger} \sim 13$–15 kcal/mol were found for those compounds having one unsubstituted benzo bridge; for those with two benzo bridges, $\Delta G^{\ddagger}$ was less than 9 kcal/mol. The low energy barriers in the 9-aryltriptycenes relative to the 9-alkyltriptycenes above were attributed to increased ground state energies in the aryl compounds.

Hindered rotation in ortho–ortho' disubstituted biphenyls has been mentioned earlier in this chapter. Often such biphenyls with bulky substituents are chiral and the optical isomers can be isolated. Thus the lack of isolatability of the cis and trans isomers of peri diphenylnaphthalenes, **69**, is surprising. Measured

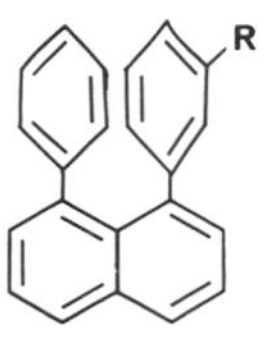

peri diphenylnaphthalenes
(69)

barriers in these compounds are low: for example, with R = CMe_2OH in **69**, $\Delta G^{\ddagger}$ = 16.4 kcal/mol, compared to the barrier $\Delta G^{\ddagger}$ = 33.5 kcal/mol in the similar but rigid [3.4]-paracyclophane, **70**.[123]

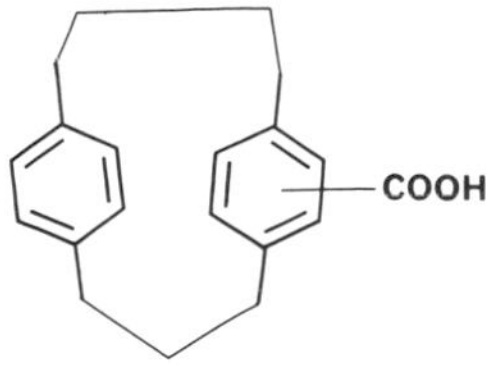

[3.4]paracyclophane
(70)

Employing ^{1}H NMR line shape analysis, Clough and Roberts[123] determined a rotational barrier of $\Delta G^{\ddagger}$ = 14.9 kcal/mol from the temperature dependences of the resonances of the diastereotopic methyl groups in the highly strained 1,4,5,8-tetraphenylnaphthalene, **71**. The low value of the activation free energy

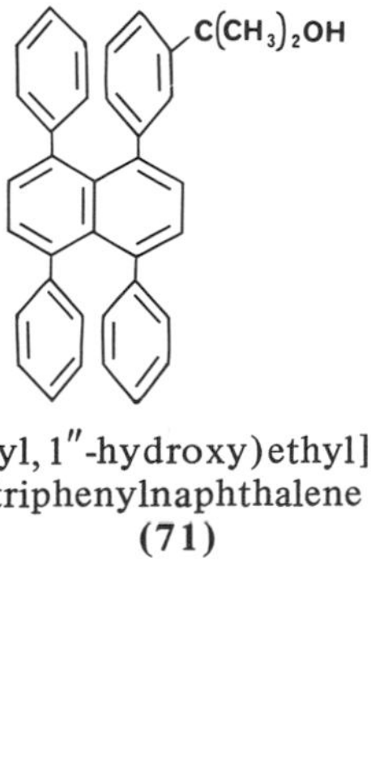

1-[3'-(1''-methyl, 1''-hydroxy)ethyl]phenyl-4,5,8-
triphenylnaphthalene
(71)

in **71** was attributed to relief of the nonbonded peri strain by large in-plane and out-of-plane distortions of the naphthalene moiety. This rationale is supported by X-ray results on 1,8-diphenylnaphthalene, **72**, which indicate the C-1–C-9–C-8

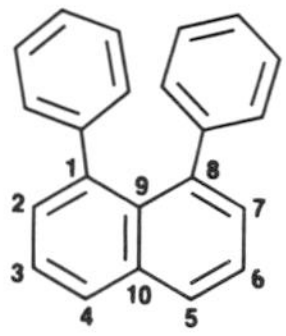

1,8-diphenylnaphthalene
(72)

angle is opened to ~126°, the C-9–C-1–phenyl angle is opened to about 125°, and the naphthyl–phenyl out-of-plane angles are about 2°. Thus the flexibility of the naphthalene nucleus lowers the barrier relative to those in the cyclophanes, which lack phenyl–phenyl splaying. A similar mechanism is proposed for other peri-substituted naphthalenes.

Related studies of other phenyl-substituted aromatic compounds, including phenylanthracenes, have been performed.[124] Rotational barriers, measured from the coalescence of ortho and meta ^{13}C chemical shifts, ranged from 14 to more than 18 kcal/mol.

Slow rotation about the naphthyl–imino bond in N-[2-methyl-1-(1-naphthyl) propylidene]benzylamine, **73**, is evidenced by a variety of effects in the variable-temperature ^{1}H spectra of this compound.[125] These include geminal anisochronism of the methyl groups, which are diastereotopic because of the hindered

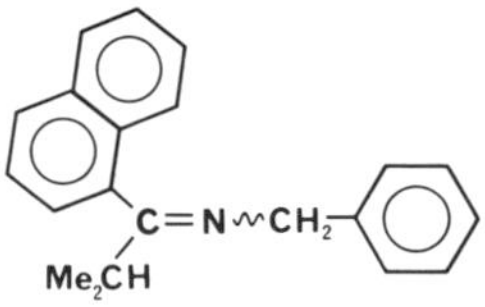

N-[2-methyl-1-(1-naphthyl)
propylidene]benzylamine
(73)

rotation. Unlike the ^{1}H spectra, the ^{13}C spectra did not show anisochronism of the geminal methyl carbons. Thus the results suggest that caution should be exercised in drawing conclusions from variable-temperature studies on geminal nonequivalence.

Hindered Rotation About Partial Double Bonds

Rotational impedance resulting from the partial double-bond character of single bonds between aromatic moieties and groups that can participate in conjugation with the ring π-electron system is common. As in sterically hindered systems, in partially double-bonded systems the rotational barriers are often of the right size to produce line shape changes in the NMR spectra.

Drakenberg and his colleagues[126] have studied the barriers to rotation and deduced the conformations of the aldehyde group in some naphthaldehydes and azulenealdehydes. The compounds studied are 1-azulenecarbaldehyde, **74**; 1,3-azulenedicarbaldehyde, **75**; 1-acetylazulene, **76**; 1-naphthaldehyde, **77**; 2-naphthaldehyde, **78**; and several substituted naphthaldehydes. In 1-acetyl-azulene, **76**, the ^{13}C NMR spectrum at $-150°C$ exhibited only one set of signals;

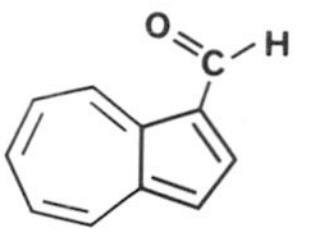

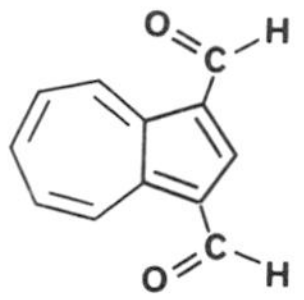

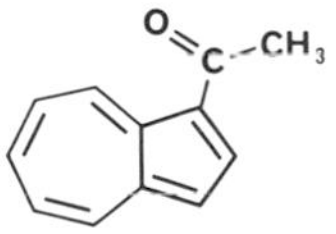

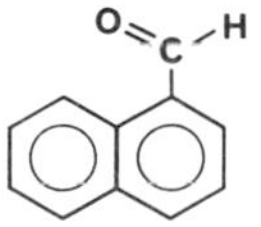

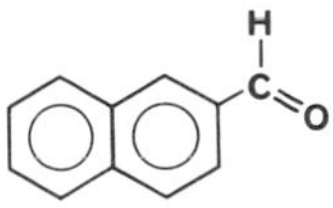

1-azulenecarbaldehyde	1,3-azulenedicarbaldehyde
(74)	**(75)**

1-acetylazulene	1-naphthaldehyde	2-naphthaldehyde
(76)	**(77)**	**(78)**

thus, in **76** either one conformer is dominant or the rotational barrier is exceedingly small. In the other aldehydes, however, ^{13}C NMR line shape analysis allowed determination of the rotational barriers. For the aldehydes **74, 77**, and **78**, the $\Delta G^{\ddagger}$ values are 42.7, 26.8, and 34.4 kJ/mol, respectively. The relative ordering of the $\Delta G^{\ddagger}$ values is in agreement with that predicted by CNDO/2 calculations.

These authors were also able to deduce the conformations of the dominant rotamers from the carbon chemical shifts. The Z rotamer (with the carbonyl group trans to the naphthalene C-1,C-2 bond, in **79**) was found to be dominant in **74** and **77**, whereas the E rotamer, **80**, is dominant in **78**. The effects of various substituents on the rotational barriers were also measured.

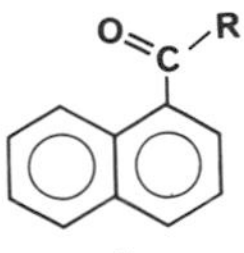

Z rotamer of 1-naphthaldehyde	E rotamer of 2-naphthaldehyde
(79)	**(80)**

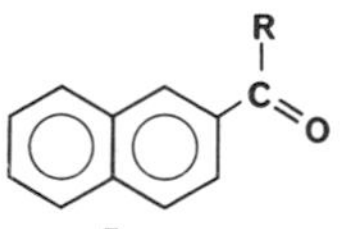

Conjugation of the nitrogen lone pair electrons with the aromatic ring in the N,N-dimethylpyrylium salt, **81**, leads to partial π character and hence restricted

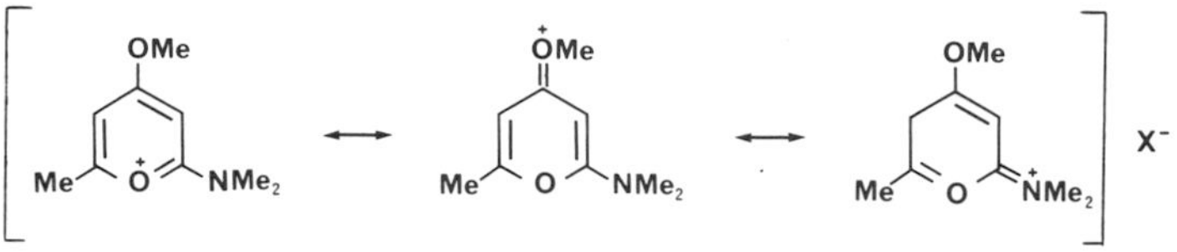

2-N,N-dimethylamino-4-methoxy-6-methylpyrylium salts
(81)

rotation about the N-aryl bond. ^{13}C NMR line shape analysis allowed extraction of the barriers to rotation about this partial double bond.[127] As the 4-substituent in **81** is changed from phenyl to methyl to methoxy to dimethylamino, the $\Delta G^{\ddagger}$ values decrease from 19.1 kcal/mol (79.9 kJ/mol) to 12.6 kcal/mol (52.7 kJ/mol). Thus the rotational barrier at C-2,N decreases with increasing electron-donating capability of the substituent at C-4.

In protonated N,N-dimethylamino-4-pyrimidines, **82**, conjugation between the dimethylamino group and the ring gives rise to rotational hindrance about

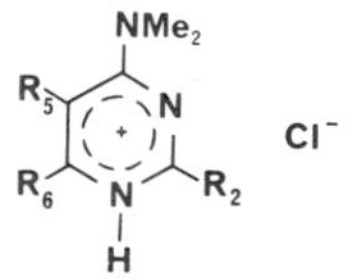

N,N-dimethylamino
pyrimidine hydrochlorides
(82)

the C-4,N bond.[128] Proton and carbon NMR line shape analysis shows the barrier in the nonprotonated species to be $\Delta G^{\ddagger}$ = 14 kJ/mol, whereas in the monoprotonated species, $\Delta G^{\ddagger}$ = 24 kJ/mol. The degree of conjugation is thus greater in the ion than in the neutral molecule. A greater increase in $\Delta G^{\ddagger}$ from the molecule to the ion was observed for 4-dimetylamino than for 2-dimethyl-amino substitution.

Partial double-bond character may also be an attribute of N-N bonds in potentially conjugated systems. The rotational barriers about such bonds in N-nitrosocarbazole, **83**, and N-nitrosodiphenylamine, **84**, in dimethyl sulfoxide

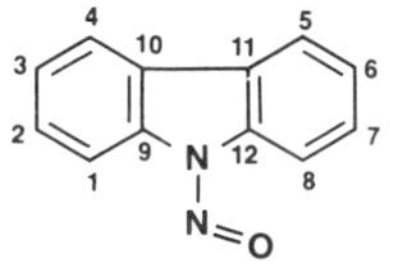

N-nitrosocarbazole
(83)

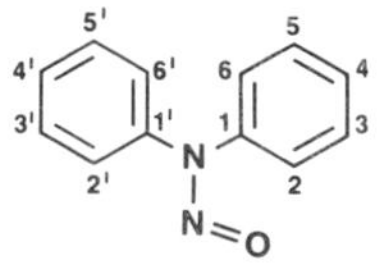

N-nitrosodiphenylamine
(84)

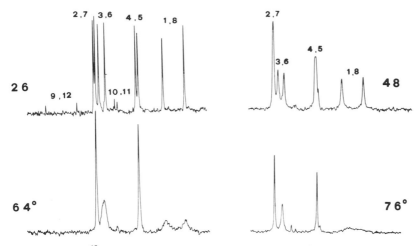

Figure 6.11. The ^{13}C NMR spectrum of *N*-nitrosocarbazole, **83**, at various temperatures in dimethyl-d_6 sulfoxide. (From reference 129.)

solution have been determined by ^{13}C NMR line shape and coalescence temperature measurements.[129] Variable-temperature ^{13}C spectra of **83** are shown in Figure 6.11. At ambient temperature, **83** manifests 12 carbon resonances, indicating a relatively high rotational barrier. Line shape analysis of these spectra yields a rotational barrier of $\Delta G^{\ddagger} = 16.85 \pm 0.5$ kcal/mol. The rotational barrier for **84**, from coalescence lifetimes, is $\Delta G^{\ddagger} = 19.1 \pm 0.1$ kcal/mol; for its 4,4'-dibromo derivative, $\Delta G^{\ddagger} = 18.8 \pm 0.15$ kcal/mol. In **83**, the lower barrier may be a result of better coplanarity of the more rigid system. As expected, these aromatic nitrosamines exhibit lower rotational barriers than their aliphatic counterparts, which typically have $\Delta G^{\ddagger}$ values around 23 kcal/mol and have greater contributions from mesomeric forms $>N^+ = N\text{-}O^-$ than do the aromatic nitrosamines. Interestingly, earlier studies of **84** in methylene chloride from -92 to 40°C showed no evidence of hindered rotation.[130] Stabilization of the mesomeric forms above by the dimethyl sulfoxide undoubtedly accounts for the higher barrier in this solvent.

REFERENCES

1. J. T. Arnold, *Phys. Rev.* **102**, 136 (1956).

2. H. S. Gutowsky and C. H. Holm, *J. Chem. Phys.* **25**, 1228 (1956).

3. E. D. Becker, *High Resolution NMR: Theory and Chemical Applications*, 2nd ed., Academic, New York, 1980.

4. R. J. Abraham and P. Loftus, *Proton and Carbon-13 NMR Spectroscopy: An Integrated Approach*, Heyden, London, 1978.

5. L. M. Jackman and F. A. Cotton (Eds.): *Dynamic Nuclear Magnetic Resonance Spectroscopy*, Academic, New York, 1975.

6. L. M. Jackman and S. Sternhell, *Applications of Nuclear Magnetic Resonance Spectroscopy in Organic Chemistry*, 2nd ed., Pergamon, Oxford, 1969.

7. S. Szymanski, M. Witanowski, and A. Gryff-Keller, *Annu. Rep. NMR Spectrosc.* **8**, 227 (1978).

8. A. Steigel, *NMR: Basic Princ. Prog.* **15**, 1 (1978).

9. H. S. Gutowsky, "Time-Dependent Magnetic Perturbations" *in* L. M. Jackman and F. A. Cotton (Eds.): *Dynamic Nuclear Magnetic Resonance Spectroscopy*, pp. 1–21, Academic, New York, 1975.

10. E. Breitmaier and W. Voelter, ^{13}C *NMR Spectroscopy*, 2nd ed., Verlag Chemie, New York, 1978.

11. B. E. Mann, *Prog. NMR Spectrosc.* **11**, 95 (1977).

12. N. K. Wilson and J. B. Stothers, *Top. Stereochem.* **8**, 1 (1974).

12a. J. B. Lambert, R. J. Nienhuis, and J. W. Keepers, *Angew. Chem. Int. Ed. Engl.* **20**, 487 (1981).

13. J. B. Stothers, *Carbon-13 NMR Spectroscopy*, Academic, New York, 1972.

14. G. C. Levy and G. L. Nelson, *Carbon-13 Nuclear Magnetic Resonance for Organic Chemists*, Wiley-Interscience, New York, 1972.

15. D. A. Wright, D. E. Axelson, and G. C. Levy, "Physical Chemical Applications of ^{13}C Spin Relaxation Measurements" *in* G. C. Levy (Ed.): *Topics in Carbon-13 NMR Spectroscopy*, Vol. 3, pp. 103–284, Wiley-Interscience, New York, 1979.

16. H. W. Spiess, *NMR: Basic Princ. Prog.* **15**, 55 (1978).

17. M. Holz and M. D. Feidler, *Prog. Nucl. Magn. Resonance* **6**, 92 (1977).

18. R. E. London and J. Avitabile, *J. Chem. Phys.* **65**, 2443 (1976).

19. F. W. Wehrli, "Organic Structure Assignments Using ^{13}C Spin–Relaxation Data" *in* G. C. Levy (Ed.): *Topics in Carbon-13 NMR Spectroscopy*, Vol. 2, pp. 343–389, Wiley-Interscience, New York, 1976.

20. R. Freeman and H. D. W. Hill, "Determination of Spin–Spin Relaxation Times in High-Resolution NMR," *in* Ref. 5, pp. 131–162.

21. J. R. Lyerla, Jr. and G. C. Levy, "Carbon-13 Nuclear Spin Relaxation" *in* G. C. Levy (Ed.): *Topics in Carbon-13 NMR Spectroscopy*, Vol. 1, pp. 79–148, Wiley-Interscience, New York, 1974.

22. W. G. Klemperer, "Delineation of Nuclear Exchange Processes," *in* Ref. 5, pp. 23–44.

23. L. W. Reeves, "Application of Nonselective Pulsed NMR Experiments—Diffusion and Chemical Exchange," *in* Ref. 5, pp. 83–130.

24. G. Binsch, "Band-Shape Analysis," *in* Ref. 5, pp. 45–81.

25. G. Binsch, *Top. Stereochem.* **3**, 97 (1968).

26. W. B. Moniz, C. F. Poranski, Jr., and S. A. Sojka, "^{13}C CIDNP as a Mechanistic and Kinetic Probe" in G. C. Levy (Ed.): *Topics in Carbon-13 NMR Spectroscopy*, Vol. 3, pp. 361–389, Wiley-Interscience, New York, 1979.

27. J. B. Stothers, "^{13}C NMR Studies of Reaction Mechanisms and Reactive Intermediates," *in* G. C. Levy (Ed.): *Topics in Carbon-13 NMR Spectroscopy*, Vol. 3, pp. 229–286, Wiley-Interscience, New York, 1979.

28. L. A. Telkowski and M. Saunders, "Dynamic NMR Studies of Carbonium Ion Rearrangements," *in* Ref. 5, pp. 523–541.

29. E. Grunwald and E. K. Ralph, "Proton Transfer Processes," *in* Ref. 5, pp. 621–647.

30. C. A. Fyfe, M. Cocivera, and S. W. H. Damji, *Acc. Chem. Res.* **1978**, 277.

31. P. E. Hansen, *Org. Magn. Resonance* **12**, 109 (1979).

32. C. P. Slichter, *Principles of Magnetic Resonance*, Harper and Row, New York, 1963.

33. A. Carrington and A. D. McLachlan, *Introduction to Magnetic Resonance*, Harper and Row, New York, 1967.

34. N. K. Wilson, "Carbon-13 NMR Chemical Shifts and Spin-Lattice Relaxation Times of Terphenyls," paper presented at the 13th Southeastern Magnetic Resonance Conference, Durham, N.C., October, 1981. Abstract D-2.

35. A. Allerhand and D. Doddrell, *J. Am. Chem. Soc.* **93**, 2777 (1971).

36. A. Allerhand, D. Doddrell, and R. Komoroski, *J. Chem. Phys.* **55**, 189 (1971).

37. J. H. Noggle and R. E. Schirmer, *The Nuclear Overhauser Effect*, Academic, New York, 1971.

38. T. D. Alger, W. D. Hamill, Jr., R. J. Pugmire, D. M. Grant, G. D. Silcox, and M. Solum, *J. Phys. Chem.* **84**, 632 (1980).

39. V. G. Berezhnoi and N. M. Sergeev, *Zh. Strukt. Khim.* **16**, 136 (1975).

40. W. A. Anderson, R. Freeman, and H. Hill, *Pure Appl. Chem.* **32**, 27 (1972).

41. W. Kitching, M. Bullpitt, D. Gartshore, W. Adcock, T. C. Khor, D. Doddrell, and I. D. Rae, *J. Org. Chem.* **42**, 2411 (1977).

42. I. Morishima, K. Kawakami, T. Yonezawa, K. Goto, and M. Imanari, *J. Am. Chem. Soc.* **94**, 6555 (1972).

43. R. S. Ozubko, G. W. Buchanan, and I. C. P. Smith, *Can. J. Chem.* **52**, 2493 (1974).

44. J. N. Shoolery, *Prog. Nucl. Magn. Resonance Spectrosc.* **11**, 79 (1977).

45. R. E. Wasylishen, B. A. Pettit, and W. Danchura, *Can. J. Chem.* **55**, 3602 (1977).

46. N. K. Wilson and R. D. Zehr, *J. Org. Chem.* **43**, 1768 (1978).

47. T. D. Alger, D. M. Grant, and R. K. Harris, *J. Phys. Chem.* **76**, 281 (1972).

48. S. Berger, F. R. Kreissl, D. M. Grant, and J. D. Roberts, *J. Am. Chem. Soc.* **97**, 1805 (1975).

49. J. Goulon, D. Canet, M. Evans, and G. J. Davies, *Mol. Phys.* **30**, 973 (1975).

50. K. F. Kuhlmann and D. M. Grant, *J. Chem. Phys.* **55**, 2998 (1971).

51. H. Nakanishi and O. Yamamoto, *Chem. Phys. Lett.* **35**, 407 (1975).

52. D. K. Dalling, K. H. Ladner, D. M. Grant, and W. R. Woolfenden, *J. Am. Chem. Soc.* **99**, 7142 (1977).

53. R. Freeman and H. D. W. Hill, *J. Chem. Phys.* **55**, 1985 (1971).

54. R. R. Shoup and D. L. VanderHart, *J. Am. Chem. Soc.* **93**, 2053 (1971).

55. N. Boden, *in* F. C. Nachod and J. J. Zuckerman (Eds.): *Determination of Organic Structures by Physical Methods*, Vol. 4, p. 51, Academic, New York, 1971.

56. G. C. Levy, *J. Chem. Soc. Chem. Commun.* **1972**, 47.

57. G. C. Levy, J. D. Cargioli, and F. A. L. Anet, *J. Am. Chem. Soc.* **95**, 1527 (1973).

58. R. J. Highet and J. M. Edwards, *J. Magn. Resonance* **17**, 336 (1975).

59. R. K. Harris and R. H. Newman, *Mol. Phys.* **38**, 1315 (1979).

60. D. E. Woessner, *J. Chem. Phys.* **37**, 647 (1962).

61. E. J. Pedersen, R. R. Vold, and R. L. Vold, *Mol. Phys.* **35**, 997 (1978).

62. G. C. Levy, J. D. Cargioli, and F. A. L. Anet, *J. Am. Chem. Soc.* **95**, 1527 (1973).

63. N. K. Wilson, *J. Am. Chem. Soc.* **97**, 3573 (1975).

64. S. Lötjönen and P. Äyräs, *Finn. Chem. Lett.* **1978**, 260.

65. G. C. Levy, *Tetrahedron Lett.* **1972**, 3709.

66. S. Forsén and R. A. Hoffman, *Acta Chem. Scand.* **17**, 1787 (1963).

67. S. Forsén and R. A. Hoffman, *J. Chem. Phys.* **39**, 2982 (1963).

68. S. Forsén and R. A. Hoffman, *J. Chem. Phys.* **40**, 1189 (1964).

69. S. Forsén and R. A. Hoffman, *Prog. NMR Spectrosc.* **1**, 15 (1966).

70. J. B. Lambert and J. W. Keepers, *J. Magn. Resonance* **38**, 233 (1980).

71. I. D. Campbell, C. M. Dobson, R. G. Ratcliffe, and R. J. P. Williams, *J. Magn. Resonance* **29**, 397 (1978).

72. R. S. Norton, R. P. Gregson, and R. J. Quinn, *J. Chem. Soc. Chem. Commun.* **1980**, 339.

73. J. Jen, *J. Magn. Resonance* **30**, 111 (1978).

74. D. M. Doddrell, M. R. Bendall, P. F. Barron, and D. T. Pegg, *J. Chem. Soc. Chem. Comm.* **1979**, 77.

75. F. R. Jensen, C. H. Bushweller, and B. H. Beck, *J. Am. Chem. Soc.* **91**, 344 (1969).

76. C. S. Johnson, Jr., *Adv. Magn. Resonance* **1**, 33 (1965).

77. J. Feeney, G. A. Newman, and P. J. S. Pauwels, *J. Chem. Soc. (C)* **1970**, 1842.

78. P. Bouchet, A. Fruchier, G. Joncheray, and J. Elguero, *Org. Magn. Resonance* **9**, 716 (1977).

79. A. Fruchier, J. Elguero, A. F. Hegarty, and D. G. McCarthy, *Org. Magn. Resonance*, **13**, 339 (1980).

80. J. Elguero, C. Marzin, A. R. Katritzky, and P. Linda, *The Tautomerism of Heterocycles*, p. 489, Academic, New York, 1976.

81. J. Daunis, M. Follet, and C. Marzin, *Org. Magn. Resonance* **13**, 330 (1980).

82. J. Elguero, C. Marzin, and J. D. Roberts, *J. Org. Chem.* **39**, 357 (1974).

83. M. T. Chenon, C. Coupry, D. M. Grant, and R. J. Pugmire, *J. Org. Chem.* **42**, 659 (1977).

84. G. M. Coppola, R. Damon, A. D. Kahle, and M. J. Shapiro, *J. Org. Chem.* **46**, 1221 (1981).

85. P. E. Pfeffer, K. M. Valentine, and F. W. Passuh, *J. Am. Chem. Soc.* **101**, 1265 (1979).

86. A. Fruchier, E. Alcade, and J. Elguero, *Org. Magn. Resonance* **9**, 235 (1977).

87. M. T. Chenon, R. J. Pugmire, D. M. Grant, R. P. Panzica, and L. B. Townsend, *J. Am. Chem. Soc.* **97**, 4636 (1975).

88. M. C. Thorpe, W. C. Coburn, Jr., and J. A. Montgomery, *J. Magn. Resonance* **15**, 98 (1974).

89. U. Edlund, *Chem. Scr.* **7**, 85 (1975).

90. M. Alei, Jr., L. O. Morgan, W. E. Wageman, and T. W. Whaley, *J. Am. Chem. Soc.* **102**, 2881 (1980).

91. M. Alei, Jr., W. E. Wageman, and L. O. Morgan, *Inorg. Chem.* **17**, 3314 (1978).

92. M. Alei, Jr., L. O. Morgan, and W. E. Wageman, *Inorg. Chem.* **17**, 2288, (1978).

93. M. D. Joesten and L. J. Schaad, *Hydrogen Bonding*, Marcel Dekker, New York, 1974.

94. U. Ewers, H. Günther, and L. Jaenicke, *Chem. Ber.* **106**, 3951 (1973).

95. P. Van de Weijer, H. Thijsse, and D. Van der Meer, *Org. Magn. Resonance* **8**, 187 (1976).

96. U. Ewers, A. Gronenborn, H. Günther, and L. Jaenicke, *Chem. Biol. Pteridines, Proc. Int. Symp.* p. 687, 1975.

97. G. A. Olah, R. H. Schlosberg, R. D. Porter, Y. K. Mo, D. P. Kelly, and G. D. Mateescu, *J. Am. Chem. Soc.* **94**, 2034 (1972).

98. G. A. Olah, G. D. Mateescu, and Y. K. Mo, *J. Am. Chem. Soc.* **95**, 1865 (1973).

99. V. A. Koptyug, I. S. Isaev, and A. F. Rezvukhin, *Tetrahedron Lett.* **1967**, 823.

100. G. A. Olah, J. S. Staral, G. Asencio, G. Liang, D. A. Forsyth, and G. D. Mateescu, *J. Am. Chem. Soc.* **100**, 6299 (1978).

101. J. F. M. Oth, K. Müllen, H. Königshofen, J. Wassen, and E. Vogel, *Helv. Chim. Acta* **57**, 2387 (1974).

102. E. Vogel, J. Wassen, H. Königshofen, K. Müllen, and J. F. M. Oth, *Angew. Chem.* **86**, 777 (1974).

103. G. Scatchard, *Ann. N.Y. Acad. Sci.* **51**, 660 (1949).

104. N. K. Wilson, *J. Am. Chem. Soc.* **94**, 2431 (1972).

105. N. K. Wilson and W. E. Wilson, *Sci. Total Environ.* **1**, 245 (1972).

106. H. Fujiwara, F. Sakai, M. Takeyama, and Y. Sasaki, *Bull. Chem. Soc. Japan* **54**, 1380 (1981).

107. Y. Nosaka, K. Akasaka, and H. Hatano, *J. Phys. Chem.* **26**, 2829 (1978).

108. R. G. Griffith, D. M. Grant, and J. D. Roberts, *J. Org. Chem.* **40**, 3726 (1975).

109. Y. K. Grishin, N. M. Sergeyev, and Y. A. Ustynyuk, *Org. Magn. Resonance* **4**, 377 (1972).

110. I. Prins, J. W. Verhoeven, and T. J. deBoer, *Org. Magn. Resonance* **9**, 543 (1977).

111. N. K. Wilson, *J. Phys. Chem.* **83**, 2649 (1979).

112. K. Hayamizu, M. Murato, and O. Yamamoto, *Bull. Chem. Soc. Japan* **48**, 1842 (1975).

113. F. H. Koehler, *Chem. Ber.* **107**, 570 (1974).

114. F. Coletta, A. Gambaro, G. Rigatti, and A. Venzo, *Spectrosc. Lett.* **10**, 971 (1977).

115. D. J. Ciappenelli, F. A. Cotton, and L. Kruczynski, *J. Organomet. Chem.* **42**, 159 (1972).

116. N. M. Sergeyev, Y. K. Grishin, Y. N. Luzikov, and Y. A. Ustynyuk, *J. Organomet. Chem.* **38**, C1 (1972).

117. K. G. Orell, V. Sik, M. D. Dunster, and E. W. Abel, *J. Chem. Soc. Faraday Trans. II* **71**, 631 (1975).

118. G. A. Taylor and P. E. Rakita, *Org. Magn. Resonance* **6**, 644 (1974).

119. F. A. Cotton, B. E. Hanson, J. R. Kolb, and P. Lahuerta, *Inorg. Chem.* **16**, 89 (1977).

120. G. A. Olah, L. D. Field, M. I. Watkins, and R. Malhotra, *J. Org. Chem.* **46**, 1761 (1981).

121. H. Nakanishi and O. Yamamoto, *Bull. Chem. Soc. Japan* **51**, 1777 (1978).

122. M. Nakamura and M. Ōki, *Bull. Chem. Soc. Japan* **48**, 2106 (1975).

123. R. L. Clough and J. D. Roberts, *J. Am. Chem. Soc.* **43**, 1328 (1978).

124. A. Caspar, S. Altenburger-Combrisson, and F. Gobert, *Org. Magn. Resonance* **11**, 603 (1978).

125. W. B. Jennings, S. Al-Showiman, M. S. Tolley, and D. R. Boyd, *Org. Magn. Resonance* **9**, 151 (1977).

126. T. Drakenberg, J. Sanström, and J. Seita, *Org. Magn. Resonance* **11**, 246 (1978).

127. M. T. Chenon, S. Sib, and M. Simalty, *Org. Magn. Resonance* **12**, 71 (1979).

128. J. Riand, M. T. Chenon, and N. Lumbroso-Bader, *Can. J. Chem.* **58**, 466 (1980).

129. L. Forlani, L. Lunazzi, D. Macciantelli, and B. Minguzzi, *Tetrahedron Lett.* **1979**, 1451.

130. N. K. Wilson, unpublished data.

Seven

Experimental Techniques

Throughout this book we have emphasized the many types of chemical information that can be obtained from the nuclear magnetic resonance spectra of aromatic compounds. Detailed coverage of the numerous experimental NMR techniques developed in the past few years that allow one to acquire this information is beyond the scope of this book. Thus we shall survey the available experimental methods applicable to the high-resolution NMR of aromatic compounds and leave more detailed discussion to the several available books on modern NMR methods, among them, several devoted primarily to the theory and practice of Fourier transform NMR spectroscopy.[1-3] In addition, several texts[4-12] and review articles[13-16] on applications of NMR, particularly ^{13}C NMR, provide good discussions of experimental techniques.

THE NMR METHOD

When a magnetic nucleus, such as that of ^{13}C, is placed in a magnetic field, its angular momentum is quantized. The allowed values of the angular momentum correspond to particular orientations of the nuclear spin axis with respect to the axis of the applied magnetic field. Each of these orientations—of which there are $(2I + 1)$, where I is the nuclear spin quantum number—corresponds to an allowed energy state of the nucleus, dependent on the value of the applied field, H_0, and the nuclear magnetic moment, μ. During the time the nucleus is in the field H_0, its spin axis precesses about the field axis with a characteristic frequency, ω_0. This frequency, the Larmor frequency, is related to the magnitude of the applied field by

$$\omega_0 = \gamma H_0 \tag{7.1}$$

where γ is the magnetogyric ratio of the nuclear spin species. When radio frequency (rf) energy at this frequency is applied to the spin system, it induces transitions between the allowed spin states (energy levels) of the nuclei, with the absorption or emission of quanta of energy, $h\omega_0/2\pi$. Equation 7.1 thus represents the basic resonance condition and the heart of the NMR experiment.

Continuous-Wave NMR

From the time NMR spectroscopy first attracted the interest of chemists until the late 1960s, most chemical applications of NMR were accomplished by continuous-wave (CW) experiments. In a CW–NMR experiment, the radio frequency is swept slowly through the frequency range of interest while the applied magnetic field is held constant. Alternatively, the magnetic field is swept slowly while the rf is held constant. In either of these equivalent experiments, the resonance condition, Eq. 7.1, is met only during the time the irradiating rf is on a line of width, $\Delta\nu_{obs}$, typically about 0.5 Hz. Since the spectral width may be 1000 Hz or more, only a small fraction of the experimental time is spent actually gathering information.

Fourier-Transform NMR

A more efficient way of performing an NMR experiment is to excite all the nuclear resonances in a sample at the same time. It can be done by applying a radiofrequency pulse of duration sufficiently short that its effective bandwidth covers the entire frequency range of the spectrum.[17] The rf pulse induces a magnetization, which decays with time, in the spin system. The resultant NMR signal, the free induction decay (FID), decays likewise with the time constant T_2^*, the effective spin–spin relaxation time.

All the information that is contained in the CW spectrum—such as resonance frequencies, line shapes, line widths, and intensities—is also contained in the analogous free induction decay. In the FID, however, this information is expressed as a signal amplitude as a function of time, $f(t)$, or as a time-domain spectrum. To convert the time-domain spectrum into a conventional frequency-domain spectrum, a Fourier transformation is performed on the FID. The signal amplitude as a function of frequency, $F(\omega)$, is the Fourier transform (FT) of $f(t)$:

$$F(\omega) = \int_{-\infty}^{\infty} f(t) \exp(-i\omega t)\, dt \qquad (7.2)$$

A typical FID and its Fourier transform are shown in Figure 7.1.

Fourier transform NMR has several significant advantages over slow-passage CW–NMR. Because only a few seconds are required to collect the FID, many

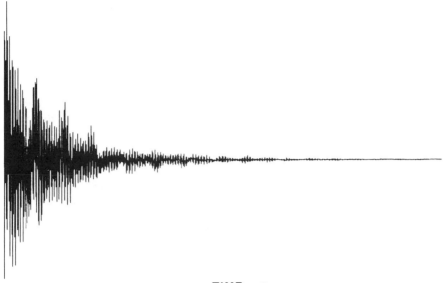

TIME $\longrightarrow$

Fourier Transformation

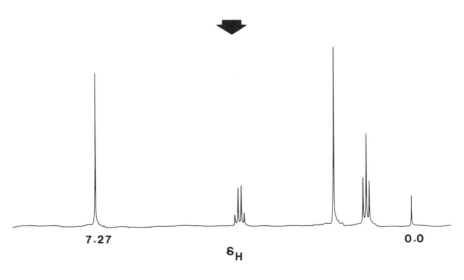

7.27

δ_H

$\longleftarrow$ FREQUENCY

0.0

Figure 7.1. The free induction decay and the conventional ^{1}H NMR spectrum obtained by Fourier transformation of the free induction decay, for a 3:1 mixture of ethyl acetate and benzene in benzene-d_6 with a tetramethylsilane internal reference.

individual spectra may be added to increase the sensitivity of the experiment. Since the signals add coherently, whereas the noise adds incoherently, the signal-to-noise ratio, S/N, increases as the square root of the number of scans. Sensitivity increases by a factor of 100 or more are possible in the same total experiment time, or alternatively, equivalent savings in experiment time can be gained.

Besides these gains in sensitivity, the Fourier transform method provides inherently better resolution than does the CW method, and this increased resolution can be enhanced further by the judicious application of weighting functions to the FID and other computer treatments of the data. Further enhancement of the S/N is also possible.

Pulsed NMR techniques and Fourier transformation allow a variety of experiments on a typical commercial NMR instrument—for example, measurements of spin–lattice relaxation times. Other applications can be accomplished with specialized pulse sequences.[1,2]

PULSED NMR EXPERIMENTS

Single-Pulse Experiments

Fourier Transform Spectra

The advent of small laboratory computers and the development of computer procedures for the rapid calculation of Fourier transforms in the last 10 years or so have led to the introduction of several different NMR methods based on FT techniques. These methods differ in the way the magnetic nuclei in the sample are excited. Pulse excitation, which we discussed briefly in the preceding section of this chapter, employs a short single-frequency rf pulse that accomplishes simultaneous excitation of all the nuclei in the sample. Such simultaneous excitation can also be accomplished by use of random or pseudo-random noise containing the entire range of frequencies of the nuclei, in the method of stochastic excitation.[18,19] Another method, rapid-scan correlation spectroscopy,[20,21] is basically a CW method employing a very fast spectral scan rate and mathematical techniques to extract the true slow-passage CW spectrum from the distorted spectrum that results. Both stochastic excitation and rapid-scan correlation NMR are much less widely used than is pulse FT–NMR, so we will now focus our attention on pulse techniques.

A typical single-pulse FT–NMR experiment is diagrammed in Figure 7.2. It consists of a short, high-power rf pulse, typically lasting from 5 to 50 μsec, which is equivalent to an rf field, $\mathbf{H}_1$, at the Larmor frequency ω_0 and which tips the macroscopic magnetization vector, $\mathbf{M}$, away from its equilibrium value, $\mathbf{M}_0$. Immediately after the pulse, $\mathbf{M}$ starts to decay back to $\mathbf{M}_0$, as is shown in

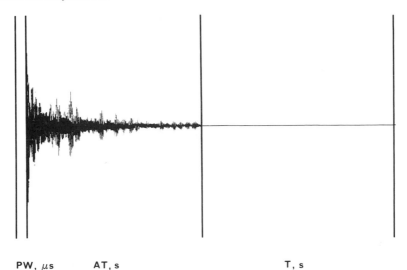

PW, μs   AT, s   T, s

Figure 7.2. A single-pulse Fourier transform NMR experiment. The pulse of length *PW* microseconds is followed by data acquisition of length *AT* seconds and a delay of *T* seconds.

Figure 7.3. The signal detected is the component of **M** in the *xy* plane, $\mathbf{M}_{xy}$, which decays to zero by spin–spin relaxation, described by T_2^*. Data acquisition occurs during this phase, over a period of from a few tenths of a second to a few seconds. In the usual case the experiment is repeated after a waiting period, *T* seconds, which allows the *z* component of the magnetization, $\mathbf{M}_z$, to return to its equilibrium value, $\mathbf{M}_0$, by spin–lattice relaxation, described by T_1, and the results of many repetitions are added to improve sensitivity.

The FID thus obtained is subjected to various weighting functions, is Fourier-transformed, and after phase corrections are applied, results in an NMR spec-

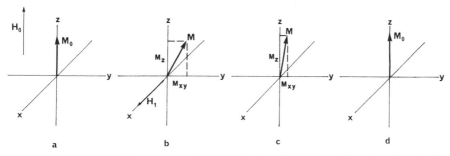

Figure 7.3. Behavior of the macroscopic magnetization, **M,** in the rotating frame in a single pulse experiment. (*a*) The equilibrium magnetization $\mathbf{M} = \mathbf{M}_0$. (*b*) **M** is tipped by action of the rf field, $\mathbf{H}_1$. (*c*) **M** returns to its equilibrium value as $\mathbf{M}_{xy}$ returns to zero and $\mathbf{M}_z$ returns to $\mathbf{M}_0$ by relaxation. (*d*) $\mathbf{M}_0$ is again established.

trum. Details of the computer and instrumental requirements can be found in several texts[1-5,11] and review articles,[13,14,22-24] two of the most useful being texts by Farrar and Becker[1] and by Shaw[2].

Multiple-Pulse Experiments

The simplest multiple-pulse experiment is, of course, the repeated single-pulse experiment described above, in which all the pulses are equally spaced and of the same length, and which produces a conventional NMR spectrum containing such information as chemical shifts, coupling constants, and line shapes. Additional valuable information, including relaxation times and nuclear Overhauser enhancements (NOEs), is accessible by the use of sequences of pulses having different lengths and different spacings. Sensitivity increases beyond those achieved by the use of simple FT versus CW techniques can sometimes be obtained through specialized pulse sequences in such experiments as DEFT (driven equilibrium Fourier transformation).[2,25] However, the greatest use of multiple-pulse excitation has been in the measurement of relaxation times.

Relaxation Time Measurements

The significance of the time constants T_1 for spin–lattice relaxation and T_2 for spin–spin relaxation, and the utility of their measured values have been discussed in Chapter 6. Here we shall describe the types of experiments from which relaxation times can be obtained.

Spin–Lattice Relaxation Times. One of the oldest means of measuring rates of spin–lattice relaxation, or T_1 values, is the adiabatic rapid passage experiment, a CW technique in which the resonance is swept very rapidly at high power, so that the spin populations are inverted. At various delay times, τ, later, the signal is recorded at a normal sweep rate and power, thus allowing measurement of the return of the magnetization, $\mathbf{M}$, to its equilibrium value, $\mathbf{M_0}$, as a function of τ. The difficulties associated with performing this experiment and the complexities of interpreting proton relaxation times limited its use considerably. With the coming of age of high-resolution pulsed Fourier transform NMR and its application to ^{13}C nuclei, whose relaxation data can often be interpreted readily in terms of chemically significant molecular properties (see Chapter 6), the much simpler pulse methods for T_1 measurements[26] came into widespread use.

There are basically three pulse FT methods in general use for measuring T_1 values—inversion recovery, progressive saturation, and saturation recovery. Each has some distinct advantages and disadvantages.[2] The exact method of choice depends to a large extent on the range of T_1 values to be measured and on the available instrumentation.

Inversion recovery is probably the most generally useful of the three methods. In this method, a basic $180°$-τ-$90°$ pulse sequence is used. The $180°$ pulse inverts the spin population so that **M** is aligned along the negative z axis. After a delay, τ, during which the magnetization relaxes partially back to its equilibrium value, M_0, a sampling $90°$ pulse results in a free induction decay, which is collected and which can be transformed into a spectrum. This sequence is diagrammed in Figure 7.4. Usually, more than one transient (FID) is acquired by repeating the pulse sequence; the transients are separated by a waiting period, T, to allow the magnetization to recover its equilibrium value. Thus the inversion recovery T_1 experiment is essentially

$$(180°\text{-}\tau\text{-}90°\text{-}T)_n \tag{7.3}$$

The signal intensity, S, as a function of τ can be expressed as

$$S = S_0 \left[1 - 2 \exp \left(\frac{-\tau}{T_1} \right) \right] \tag{7.4}$$

where S_0 is the signal intensity at equilibrium—that is, at $\tau \gg T_1$. This equation can be solved directly for T_1 by a least squares exponential fitting routine or can be rewritten to give a linear semilog plot of the data $(S_0 - S)$ versus τ,

$$\ln(S_0 - S) = \frac{-\tau}{T_1} + \ln 2S_0 \tag{7.5}$$

from which T_1 can be obtained from the slope $-1/T_1$. A stacked plot of the ^{13}C spectra of fluorenone, **1**, obtained in an inversion recovery Fourier transform experiment[27] is shown in Figure 7.5.

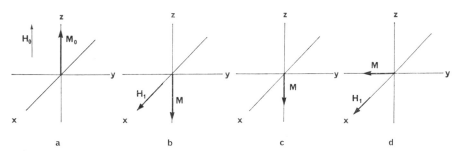

Figure 7.4. Behavior of the macroscopic magnetization, **M,** in the rotating frame in an inversion recovery T_1 experiment. (*a*) The equilibrium magnetization $M = M_0$. (*b*) The $180°$ pulse inverts **M.** (*c*) **M** decays toward M_0 by spin–lattice relaxation. (*d*) The $90°$ pulse aligns **M** along the y axis, producing a detectable signal.

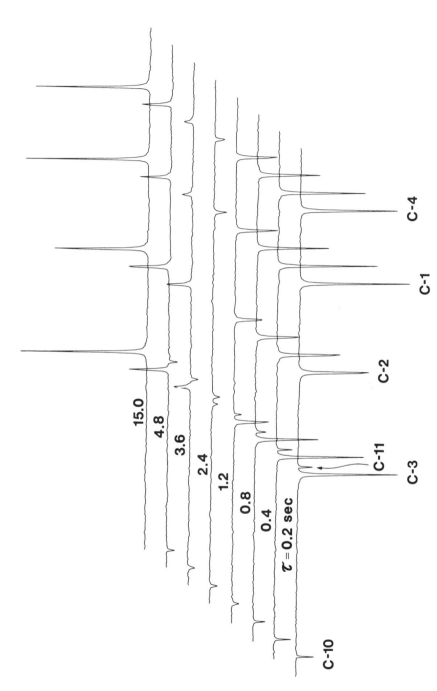

Figure 7.5. ^{13}C spectra of fluorenone, **1**, obtained in an inversion recovery measurement of T_1.

fluorenone

(1)

Several modifications of the inversion recovery sequence are possible;[2,3,28-32] these offer some advantages in accuracy or experimental time requirements.[33,34]

A second method of measuring T_1 is known as progressive saturation. In the repeated single-pulse experiment discussed earlier in this chapter, if the delay between transients is too short, the magnetization does not have a chance to decay back to equilibrium between pulses by spin–lattice relaxation; hence the signal intensity is decreased. The exact fraction by which this saturation decreases the signal depends upon the delay, τ, and upon T_1. This dependence is the basis of the progressive saturation T_1 measurement

$$(90°\text{-}\tau\text{-}90°\text{-}\tau)_n \tag{7.6}$$

Data acquisition begins after a series of a few pulses has established a steady state. The signal intensity as a function of τ is

$$S = S_0 \left[1 - \exp\left(\frac{-\tau}{T_1}\right)\right] \tag{7.7}$$

and T_1 can be extracted from the data in the same ways as in the inversion recovery experiment. A drawback to this method is inherent in the necessity of including the time required to record the free induction decay in the delay, τ, which restricts the usefulness of progressive saturation to measurement of fairly large T_1 values.

Instead of looking at the signal after a degree of saturation has been induced, one can look at the signal after a degree of recovery from saturation has taken place, since the latter process too is governed by spin–lattice relaxation.[35,36] The most useful way is to use a pulse sequence including a homogeneity spoiling pulse:

$$(90°\text{-}HS\text{-}\tau\text{-}90°\text{-}T\text{-}HS)_n \tag{7.8}$$

The homospoil pulse destroys the coherence of the spins so that there is no longer any net magnetization along any axis; the recovery of the magnetization is sampled with a second 90° pulse at time τ after the first 90° pulse, and data are acquired during time T. No additional delays are necessary, since the second homospoil pulse destroys any residual transverse magnetization. The signal intensity as a function of τ is described by the same equation as in progressive saturation, Eq. 7.7, and data analysis is accomplished in the same way. Unlike progressive saturation, however, saturation recovery is applicable to the measurement of short, as well as long relaxation times. A stacked plot of the fluorenone ^{13}C spectra obtained by this method of T_1 measurement is shown in Figure 7.6.

Detailed comparisons of all three of the above techniques for T_1 measurement and discussions of their optimum use and accuracy are in the literature.[2,34,37-40]

Spin–Spin Relaxation Times. Measurement of T_2 values, or spin–spin relaxation times, is difficult. The difficulty arises largely from the sensitivity of the relaxation process to magnetic field fluctuations at low frequencies. Thus contributions to T_2 can arise from any low-frequency processes, such as chemical exchange and diffusion, as we mentioned in Chapter 6, as well as from inhomogeneities in the applied magnetic field. The free induction decay is an example. It is normally characterized by an effective spin–lattice relaxation time, T_2^*, with

$$\frac{1}{T_2^*} = \frac{1}{T_2} + \frac{\gamma \Delta H_0}{2} \tag{7.9}$$

where ΔH_0 is the magnetic field inhomogeneity, generally greater than 0.05 Hz. If additional low-frequency processes are occurring in the sample, there will be additional terms in Eq. 7.9 that contribute to the resultant line broadening.

Various spin–echo techniques have been used to measure T_2 values, the classical experiment being that of Carr and Purcell.[41] With the introduction of pulse phase shifts to eliminate rf inhomogeneity, this experiment is described by

$$90^{\circ}_x\text{-}\tau\text{-}180^{\circ}_y\text{-}2\tau\text{-}180^{\circ}_y\text{-}2\tau\text{-}180^{\circ}_y \tag{7.10}$$

the Carr–Purcell–Meiboom–Gill (CPMG) pulse sequence.[42]

For liquids of low viscosity, with currently available commercial instrumentation, the forced transitory precession technique, or "spin-locking," is a relatively painless way to measure T_2 values. It is a modification of the CPMG sequence, in which the time between the pulses is reduced to zero. A shift of the rf phase by 90° immediately following the 90° pulse causes both $\mathbf{M}$ and $\mathbf{H}_1$ to lie along the y axis. The effective magnetic field is now $\mathbf{H}_{eff} \approx \mathbf{H}_1$, the effects of inhomogeneity in $\mathbf{H}_0$ are eliminated, and the spins are locked along the y axis. The relaxation of $\mathbf{M}$ along $\mathbf{H}_1$ that occurs is characterized by a time, $T_{1\rho}$, or T_1 in the rotating frame, and for most liquids $T_{1\rho} = T_2$.

Complications arise from echo modulation in homonuclear coupled systems under high-resolution conditions, and they present a significant obstacle to T_2 measurements in coupled spin systems. This obstacle is not insurmountable, however; several means are available,[1-3] including use of J spectra.[44]

Two-Dimensional NMR

A useful technique that is just beginning to be applied widely is two-dimensional (2D) Fourier transform NMR spectroscopy.[45] This type of experiment, in

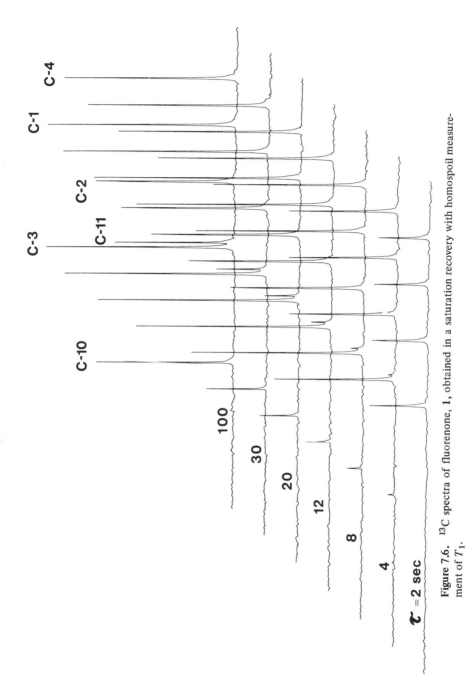

Figure 7.6. ¹³C spectra of fluorenone, **1**, obtained in a saturation recovery with homospoil measurement of T_1.

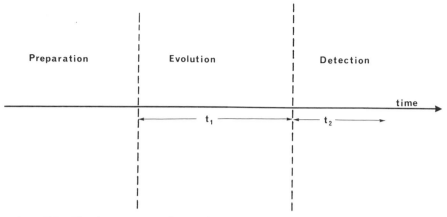

Figure 7.7. The time structure of a two-dimensional Fourier transform NMR experiment.

general form, consists of three periods, as shown in Figure 7.7. During the first period the spins are prepared in some specified state, which may be accomplished by continuous irradiation, as in broadband decoupling to establish the NOE, or a pulse, as in cross-polarization experiments[46,47] with ^{13}C. This perturbation of the spin system is followed by an evolution period, t_1, in which the spins respond to the perturbation under defined conditions. These conditions may include spin–echo refocusing as in T_2 measurements, or additional decoupling or coupling. Finally, during a detection period, t_2, the FID is collected under conditions that may differ from those during t_1 in, for example, the on/off condition of the decoupler, to allow sorting out of some of the complexities of the experiment.[48] Double Fourier transformation of a series of spectra collected with varying t_1 results in a set of spectra that depend on two frequencies. These spectra can be displayed as a contour map[49] or as a stacked plot such as the one shown in Figure 7.8.

One application of 2D Fourier transform NMR is in the determination of ^{1}H and ^{13}C spectra simultaneously. During the preparation period, the protons are irradiated with a 90° pulse, following which they precess freely during the evolution period, t_1. During t_2, 90° pulses are applied to both ^{1}H and ^{13}C nuclei, but only the ^{13}C FID is acquired. This ^{13}C FID contains not only the ^{13}C spectrum, but also information about the spin states and precession of the ^{1}H nuclei at the time of the second (detection) 90° pulse. Several spectra with different t_1 and double Fourier transformation result in complete ^{1}H and ^{13}C spectra, with automatic correlation of ^{13}C and attached proton signals. By varying the conditions, the carbon–hydrogen couplings can be displayed or removed along either frequency axis, which makes 2D Fourier transform NMR extremely useful in spectral interpretation.

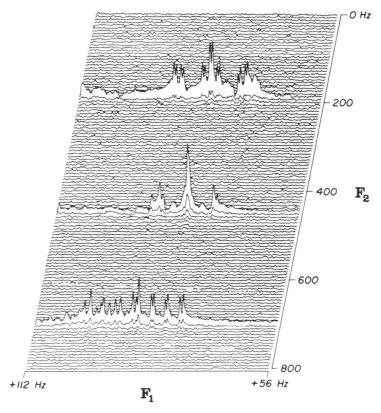

0 Hz

200

400 F_2

600

800

+112 Hz   +56 Hz

F_1

Figure 7.8. The high-frequency section of the proton flip J spectrum of pyridine, **2**. ^{13}C

chemical shifts are on the F_2 axis. The responses are for C-3, C-4, and C-2 in the upper, middle, and lower traces, respectively. (From reference 55.)

There are many other applications of 2D NMR,[16,48] including spin–echo 2D spectra (J spectra),[50-55] an example[55] of which is given in Figure 7.8, relaxation time measurements,[56] measurements of carbon–carbon spin coupling,[57] and studies of various exchange processes.[58] Because of the wealth of chemical information that can be garnered from 2D–NMR experiments, the significant investment in time necessary to carry out these experiments is often easily justified. Two-dimensional NMR capability is now incorporated in most new commercial NMR instruments. Thus we can expect to see the range and number of applications of this technique continue to grow.

Selective Excitation

In many potential applications of pulsed Fourier transform NMR, selective excitation of particular regions of the spectrum without excitation of neighboring regions (containing, for example, a strong undesirable solvent peak) would be clearly advantageous. Such selective excitation is possible using a rapid succession of identical small flip angle rf pulses, which are chosen so that the nuclei of interest are at resonance and experience a cumulative effect, whereas nuclei not at resonance precess incompletely between pulses and so produce no coherent signal.[16,59] For example, to produce an effective $\pi/2$ pulse 800 Hz from the rf, one could use a sequence of 20 $\pi/40$ pulses spaced 1.25 msec apart.

Nuclei may also be selectively excited by Fourier-synthesized, or tailored, excitation.[60,61] This means of performing the experiment is somewhat difficult experimentally but has broad potential as a complement to regular pulsed FT NMR methods.

Advances in multiple-pulse techniques are still being made, and the literature continues to report new applications and interesting innovations, such as pulse sequences, that compensate for their own imperfections.[62]

Cross-Polarization and High-Resolution NMR of Solids

High-resolution NMR spectra of solid samples have not been readily obtainable until fairly recently, largely because of the line broadening, often several thousand Hz, that results from internuclear dipolar interactions in the solid and chemical shift anisotropies. The dipolar interaction is dependent on the angle, θ, between the internuclear vector and the axis of the applied field. If the sample is spun rapidly at the "magic angle," $\theta = 54.7°$, the average dipolar interaction, $3 \cos^2 \theta - 1$, equals zero and the dipolar broadening is removed, just as it is in liquids by molecular tumbling. Magic angle spinning, while not yet available on most commercial spectrometers, can be provided through suitable probe modifications.[63]

If the nuclei of interest are rare nuclei, such as ^{13}C, ^{31}P, or ^{15}N, broadband decoupling of the abundant ^1H nuclei combined with appropriate pulsing of the irradiating and observing fields leads to transfer of the polarization of the abundant spins to the rare spins—or cross-polarization.[64] The beauty of this experiment, from the organic chemist's point of view, lies in the resultant tremendous signal enhancement of the rare spins. Thus with cross-polarization and magic angle spinning, high-resolution NMR spectra of even rare spins in solid samples may be obtained.

SAMPLE REQUIREMENTS

Quantity of Sample

The quantity of material required to obtain good quality NMR spectra in a reasonable period of time varies considerably with the sensitivity of the observed magnetic nucleus and with the instrumentation and techniques used. The NMR sensitivities of several common spin-$\frac{1}{2}$ nuclei are given in Table 7.1. For ^{1}H spectra obtained on recent model continuous wave NMR spectrometers, 10 mg or more dissolved in a few tenths of a milliliter of a nonproton-bearing solvent gives adequate sensitivity. Use of microcells, or extended signal averaging, reduces the necessary quantity to 1 mg or less. For CW spectra of less sensitive nuclei, such as ^{13}C, high concentrations (1 to 2 M) or neat liquids are required to obtain data without extended signal averaging.

Fourier transform operation extends the lower limit on sample size significantly. With FT, less than 10 μg will give a proton spectrum in a few hours;[65] this quantity can be reduced to 1 to 2 μg if microinserts for the NMR probe are used. The microinserts allow the use of standard 1.7 mm outside-diameter capillary tubes as sample tubes. The effluent from a gas chromatograph (GC) can be collected in such a capillary tube and used directly.

A major difficulty in ^{1}H NMR studies of very small samples arises from impurities present in the solvents or introduced during sample collection. Often the largest signals obtained from a microsample will be those of phthalate esters leached from plastic labware, or of the liquid phase of GC column packing. And even as little as 50 ng of absorbed HDO can give interference in a proton microsample.

The problem of impurities is not so severe with ^{13}C microsamples, but because the NMR sensitivity of ^{13}C nuclei is 1/6000 that of ^{1}H nuclei, the lower

Table 7.1. Natural Abundance NMR Sensitivities of Some Spin-$\frac{1}{2}$ Nuclei at Constant Field

Nucleus	Sensitivity
^{1}H	5675.
^{19}F	4734.
^{31}P	377.
^{13}C	1.00
^{17}O	0.0611
^{15}N	0.0209

limits on sample size are 3 to 5 mg, or approximately 100 to 200 μg if a micro-insert is used.[15] At these low levels, a patient NMR spectroscopist and an NMR instrument dedicated to looking at the same sample for four or five days are prerequisites. A booklet by Shoolery[66] is an excellent summary of microsample techniques.

When sensitivity is limited not by sample quantity, but by sample solubility, the opposite tack is taken—recourse to large-volume sample tubes and probe inserts. Some commercial spectrometers can accommodate up to 30-mm-diameter sample tubes. An accompanying disadvantage is the large volume of (usually deuterated) solvent required.

Solvents

The major requirement of an NMR solvent is that it add few or no signals of its own to the spectrum of interest. This requirement is particularly important when a large number of scans is to be accumulated; a strong solvent peak can severely limit the potential S/N enhancement by causing the dynamic range of the data system to be exceeded.

Most commerical FT–NMR spectrometers employ a deuterium internal field-frequency lock to stabilize the system, although a fluorine external lock is sometimes used. The deuterium lock is provided most simply by a deuterated solvent such as chloroform-d, acetone-d_6, or water-d_2. Use of a deuterated solvent also reduces the solvent dynamic range problem, especially for proton spectra.

Reference Materials

In most applications, a reference compound is included in the NMR sample to serve as an internal standard. For ^{13}C and ^{1}H, it is normally tetramethylsilane, TMS. The chemical shift of TMS is taken as δ = 0.0 and chemical shifts in ppm are expressed as

$$\delta = \frac{\omega - \omega_{ref}}{\omega_{ref}} \times 10^6 \tag{7.11}$$

making δ positive for nuclei that are less shielded than TMS and thus resonate at higher frequency (lower field in a field sweep experiment).

Other chemical shift reference materials are commonly used[3] for nuclei other than ^{13}C or ^{1}H. For example, ^{15}N chemical shifts can be conveniently reported relative to anhydrous liquid ammonia at 25°C,[9] but the most useful experimental reference compound for ^{15}N appears to be external nitromethane. Fluorine-19 data have been referenced to C_6F_6, CCl_3F, or C_4F_8; phosphorus-31 data to P_4O_6 or H_3PO_4 (85% in water); and oxygen-17 data to H_2O.

Suggestions have been made that all multinuclear NMR data be referenced to the same substance under the same experimental conditions. Such a universal chemical shift reference could be the proton resonance of tetramethylsilane, set at exactly 100 MHz.[70]

Related to the problem of chemical shift reference materials is the problem of correct temperature measurement inside the NMR probe, which is best accomplished by use of a chemical shift thermometer. For proton spectra, the spectra of methanol and of ethylene glycol, whose resonances have known temperature dependences,[71,72] are used for low and high temperature ranges, respectively. Several mixtures have been suggested for use in variable-temperature ^{13}C NMR experiments;[73] a mixture of 0.1 M $(C_6H_5)_3P$ and 0.1 M $(C_6H_5)_3PO$ in toluene-d_8 has been suggested for similar use in ^{31}P NMR.

ASSIGNMENT TECHNIQUES AND EXPERIMENTAL AIDS TO SPECTRAL INTERPRETATION

The main prerequisite to obtaining chemical information from NMR spectral data is spectral assignment—that is, the attribution of particular resonances, their chemical shifts, and coupling constants to particular magnetic nuclei in the sample. Although the assignment process can be quite straightforward in the case of small molecules with widely different chemical shifts, more typically it is a challenge, with much of the intellectual and experimental armament in the spectroscopist's arsenal required to defeat the complexities of spectra of polycyclic aromatic molecules.

Spectral Analysis

The process of assigning spectra of complex coupled spin systems, such as ^{1}H spectra of most aromatic molecules, is somewhat familiar to most organic chemists. Where substitution reduces the potential assignees to a small number, and where these nuclei reside in fairly different chemical environments, the resultant ^{1}H spectra often present an easily recognized pattern, such as AX, AB, ABX, or A$_2$B, from which chemical shifts and coupling constants can be derived with a minimum of computational effort.[75,76] With a large number of spins, four or more, and with coupling constants of the same order of magnitude as the chemical shifts, the problem quickly becomes more difficult. One of the most fruitful approaches to dealing with the difficulties is computer-aided iterative spectral analysis. Several programs have been developed over the years for these analyses; the most convenient is probably LAOCOÖN III.[77,78] From trial chemical shifts and coupling constants derived from preliminary examination of the spectrum of interest, a theoretical spectrum is calculated. More an art than a science, the next step is matching experimental and theoretical line frequencies. The computer then iterates to a best fit set of chemical shifts and coupling constants by minimizing the differences between observed and calculated line frequencies. If the original guesses of the NMR parameters are reasonably close to the true values, then the process gives excellent results. Modified versions of LAOCOÖN, which take line intensities as well as frequencies into account, are available.[79] Additionally, operating software for most commercial FT–NMR spectrometers includes forms of LAOCOÖN adapted to laboratory minicomputers.

Several texts describe the art of spectral analysis in some detail. The book by Abraham[76] is an especially useful guide to analyses of homonuclear coupled spectra.

Interpretation of NMR spectra of nuclei other than protons is often made easier by partial or complete analysis of the heteronuclear coupled spectra. Because one-bond ^{13}C-^1H coupling constants are large, typically 150–180 Hz in aromatic systems, proton-coupled ^{13}C NMR spectra are frequently second order, and require computer analysis to extract chemical shifts and coupling constants. Nevertheless, if a ^{13}C spectrum is sufficiently first order, the long-range coupling constants may be accessible and provide assignment information. In nonheterocyclic aromatic molecules, the geminal (two-bond) long-range coupling constant, $^2J_{CH}$, and the four-bond coupling constant, $^4J_{CH}$, are both about 1–2 Hz, whereas the vicinal (three-bond) coupling constant is much larger, about 7–12 Hz.[80] These relationships have been put to good use in, for example, assignments of the ^{13}C NMR spectra of substituted naphthalenes.[81,82] In heterocyclic aromatic compounds, $^2J_{CH}$ may also be large, which somewhat restricts the use of long-range couplings is assignment of the ^{13}C spectra of these substances.

Decoupling Techniques

Selective Decoupling

One of the classic means to unravel spectra of complex spin systems is single-frequency homonuclear decoupling. The application of rf power sufficient to saturate a given set of spins effectively removes their coupling to other nuclei, with the result that the signals from the other nuclei are simplified. If the saturated spins are the only ones with significant coupling to the observed spins, then the observed spin signal collapses to a single line. By moving the decoupling rf through the resonances in the spectra, one can gain insights into which nuclei are coupled to each other.

Similarly, single-frequency heteronuclear decoupling may be used to establish spectral assignments. In ^{13}C spectra, for example, decoupling frequencies can be set to the previously determined ^{1}H chemical shifts one by one; the ^{13}C signals that lose their doublet structure originating in their coupling with directly bonded protons can thus be correlated with the ^{1}H resonances. To irradiate only the protons in a single magnetic environment, the proton decoupler power is kept at the bare minimum necessary to collapse the corresponding carbon resonance.[82,83]

Off-resonance decoupling is another use of single-frequency heteronuclear coupling that is especially valuable in assignment of ^{13}C NMR spectra. The coherent rf off-resonance reduces the one-bond ^{13}C-^{1}H coupling to a fraction of its full value. The residual splittings, J_r, of the ^{13}C resonances depend on the frequency offset, $\Delta\nu$, of the decoupler from resonance and the decoupler power, H_2. If $\gamma H_2/2\pi \gg \Delta\nu$ and J, then the residual splittings are

$$J_r = J\Delta\nu(\gamma H_2/2\pi)^{-1} \qquad (7.12)$$

where J is the full coupling (in the absence of H_2).[84] Much of the complexity of the proton-coupled ^{13}C spectrum is eliminated: the residual splittings, J_r, are small enough so that there is usually little overlap between signals from various carbons.[85] Long-range ^{1}H-^{13}C couplings are normally removed, so that the multiplicities of the remaining splittings reflect only the directly bonded protons. Methyl, methylene, methine, and quaternary carbons are thus easily identified.

If the proton chemical shifts and the one-bond carbon–hydrogen coupling constants $^1J_{CH}$ are known, the SFORD (single frequency off-resonance decoupled) ^{13}C spectrum allows specific assignments to be made, with one experiment only. Even without exact knowledge of the δ_H and $^1J_{CH}$ values, assignment is sometimes possible, if the *ordering* of the proton chemical shifts is established. Values of $^1J_{CH}$ can be estimated,[86] and Eq. 7.12 can be used as before. Graphical methods are also available for interpretation of SFORD data.[87,88]

Broadband Decoupling

Most ^{13}C spectra are normally obtained with full random or pseudorandom noise decoupling of protons, for two basic reasons. First, the sensitivity (S/N) of the experiment is increased by as much as a factor of 3 because of nuclear Overhauser enhancements of the ^{13}C resonances in the presence of the decoupling field. Second, the elimination of multiplet structure gathers all the intensity of a given resonance into one peak, thereby further increasing the sensitivity. Rather than use random noise, it is possible to decouple the protons with an audio-modulated single frequency centered on the proton spectrum, with some additional improvement in S/N.[89]

Unfortunately, the S/N improvements obtained by broadband decoupling are usually at the expense of structural information provided by spin–spin coupling.

This need not be, however. By appropriately gating the decoupler field, the sensitivity enhancements caused by the nuclear Overhauser effect can be retained. The NOE decays exponentially with a time constant T_1 after the decoupler is turned off, whereas the rf coherence disappears at once. Thus if the decoupler is gated on during the delay period, T, with $T \gg T_1$, but off during data acquisition, one obtains a coupled spectrum with the full NOE. The latter experiment, in conjunction with a standard noise-decoupled experiment, allows NOE measurements to be made.

The capability of obtaining proton-coupled ^{13}C spectra without losing the sensitivity caused by nuclear Overhauser enhancements may enable one to use a "fingerprint" method for assignment of the ^{13}C spectra of aromatic molecules.[90,91] In general, the long-range ^{13}C–^{1}H couplings in aromatic compounds are ordered $^3J_{meta} > {}^3J_{ortho} > {}^4J_{para}$, although there are exceptions. In specific compounds, such as ortho-disubstituted benzenes, 3, quite different multiplet patterns are obtained for α and β carbons. An example,[90] the proton-coupled ^{13}C spectrum of naphthalene, 4, is shown in Figure 7.9.

o-disubstituted benzenes   naphthalene
(3)   (4)

Off-resonance *broadband* decoupling is occasionally useful for the elimination or reduction of ^{13}C NMR signals from protonated carbons, so that signals from quaternary carbons may be more easily identified. The residual broadening from incomplete broadband decoupling is proportional to the square of the ^{13}C–^{1}H

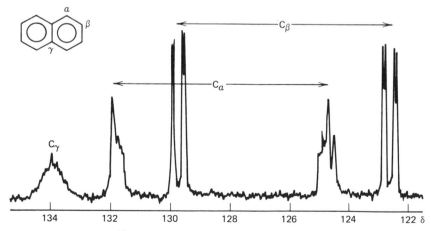

Figure 7.9. The ^{13}C NMR spectrum of naphthalene, 4. (From reference 90.)

spin–spin coupling constant.[84] Since the one-bond coupling constants $^1J_{CH}$ are larger than the long-range two- or three-bond coupling constants by at least an order of magnitude, the residual broadening of protonated carbon resonances is several orders of magnitude greater than it is for quaternary carbon resonances. This technique is especially useful for extrication of quaternary carbon resonances from the ^{13}C spectra of large molecules.

Spin–Lattice Relaxation Times

The usefulness of spin–lattice relaxation time measurements in spectral assignment rests on the dependence of T_1 on the correlation time for molecular motion, τ_c, and on the distance, r, between neighboring nuclei,

$$T_1^{-1} \propto \tau_c \sum r^{-6} \tag{7.13}$$

when the relaxation is dominated by internuclear dipolar interactions. For rigid molecules, all the nuclei have the same τ_c, and their T_1 values depend only on Σr^{-6}. If Σr^{-6} is essentially the same for all nuclei of interest, as it is for ^{13}C nuclei with directly bonded protons, which dominate the ^{13}C relaxation, then only the number, N, of directly bonded protons is important.

$$(NT_1)^{-1} \propto \tau_c \tag{7.14}$$

Anisotropic molecular motion or molecular flexibility can result in different τ_c values for different parts of a molecule. The T_1 values, then, are a guide to assignments, and are also useful in analyses of molecular motion. An example is shown in Figure 7.10, where preferred motion about the long molecular axis of 4-chlorobiphenyl, **5**, results in a shortened T_1 for the protonated carbon on that axis.[92]

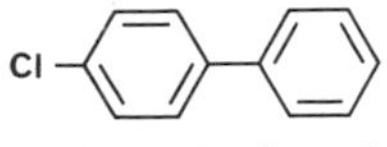

4-chlorobiphenyl
(5)

A neighboring quadrupolar nucleus, such as ^{14}N, ^{35}Cl, ^{37}Cl, ^{79}Br, or ^{81}Br, may aid assignments by scalar relaxation. Rapid fluctuations in the electric field about the quadrupolar nucleus provide dominant contributions to T_2 that make spin–spin relaxation rapid for scalar-coupled nuclei. The shortened T_2 is evidenced in the spectrum by broadened lines for nuclei directly bonded or near neighbors to quadrupolar nuclei.

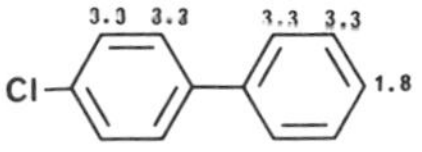

Figure 7.10. Spin–lattice relaxation times in seconds of the ^{13}C nuclei having attached protons in 4-chlorobiphenyl, **5**. (From reference 92.)

Since the nuclear Overhauser enhancements depend on the same dipole–dipole interactions as do spin–lattice relaxation times, the NOE values, too, have an r^{-6} dependence.[93] In small molecules, if a proton is irradiated, the neighboring protons, ^{13}C nuclei, and other magnetic nuclei will have resultant NMR intensity increases. This NOE can be put to good use in signal assignments, especially in assignments of the ^{13}C signals from quaternary carbons.[82,94] In 1,5-dichloronaphthalene, **6**, for example, the intensity of the C-9, -10 signal increased by a factor of 1.5 when the nearby protons H-4 and H-8 were irradiated.[82]

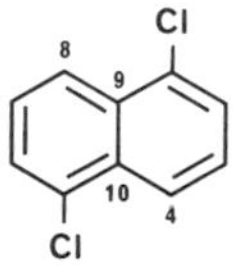

1,5-dichloronaphthalene
(6)

Relaxation and Shift Reagents

Shift Reagents

The utility of lanthanide shift reagents, such as $Eu(DPM)_3$, $Pr(DPM)_3$, and $Yb(DPM)_3$, where DPM stands for dipivalomethanato, to simplify ^1H spectra and to establish gross features of molecular structure is well known and has been thoroughly reviewed.[16,95–99] Similar utility is anticipated for ^{13}C spectra, although the number of applications currently in the literature is relatively small.

These paramagnetic rare earth complexes bind to polar groups such as OH, NH_2, or C=O in the solute of interest. The unpaired electron spins in the shift reagent induce large shifts in the resonances of the solute, primarily by a dipolar or pseudocontact mechanism. The induced shift, $\Delta H/H_0$, is then proportional to r^{-3}, the inverse cube of the distance between the rare earth ion and the given nucleus.[100]

$$\frac{\Delta H}{H_0} = Kr^{-3} (3 \cos^2 \chi - 1) \qquad (7.15)$$

The angle, χ, is the angle between the principal magnetic axis of the complex and the internuclear vector, **r**. At the critical angle $\chi = 54°44'$, $\Delta H/H_0$ changes sign, so that either upfield or downfield shifts may be induced.[101] Some typical lanthanide shifts in ^{13}C NMR spectra are given in Table 7.2.[102]

Since lanthanide-induced shifts depend on molecular geometry, they can be used in conjunction with computer calculations to establish molecular structures and stereochemistry. Thus they have been used in a determination of the conformations of several o,o,o',o'-tetrasubstituted biphenyls from their ^{1}H spectra.[103] A cautionary note is appropriate here: to determine conformations reliably from lanthanide shifts, it is necessary to know whether the shift mechanism is completely pseudocontact or whether it has contact (covalent) contributions. Contact contributions can be large, especially for europium.[104,105] Other difficulties may also present themselves.[16]

The addition of a shift reagent to a chemically exchanging species can often extend the range of rates that can be measured; the rate parameters are often more accurate as well. The use of lanthanide shift reagents in dynamic NMR studies has recently been reviewed.[106]

Simplification of the spectra of aromatic molecules with shift reagents may also allow spectral analyses with much greater accuracy; for example, a recent analysis[107] of the ^{1}H spectrum of fluorenone, **1**, is a great improvement over earlier work done without shift reagents and at lower field.[108] Silver trifluoroacetate has been used in combination with lanthanide shift reagents to provide further spectral simplification.[109]

Relaxation Reagents

Intensity ratios in ^{13}C spectra often do not reflect accurately the relative numbers of ^{13}C nuclei giving rise to the signals. This inaccuracy is a result of differences in both T_1 and NOE values for the different carbons, primarily because of their differing contributions to spin-lattice relaxation from τ_c and Σr^{-6} (see Eq. 7.13), but also occasionally from relaxation by other relaxation mechanisms. As an aid to the accurate quantitative determination of the relative numbers of carbons in different magnetic environments, relaxation reagents such as *tris*-chromium(VI)-acetylacetonate, or $Cr(acac)_3$, can be quite useful.

The paramagnetic relaxation reagent supplies an alternative relaxation pathway for the ^{13}C spins—namely, electron–nuclear relaxation—which, because the magnetic moment of the electron is about a thousand times greater than that of a proton, is much more efficient than the internuclear dipolar relaxation. The observed relaxation rate is

$$\frac{1}{T_1} = \frac{1}{T_1^{DD}} + \frac{1}{T_1^e} + \frac{1}{T_1^o} \tag{7.16}$$

where $1/T_1^{DD}$, $1/T_1^e$, and $1/T_1^o$ are the relaxation rates resulting from internuclear–dipolar, electron–nuclear, and other relaxation mechanisms, respectively. The electron–nuclear interaction normally dominates the relaxation, and in the absence of specific interactions between the relaxation reagent and the solute, causes the carbon relaxation times to become uniformly short, while quenching the nuclear Overhauser enhancements.[110] Quantitative correlations between the ^{13}C signal intensities and the number of ^{13}C nuclei then become possible.

Table 7.2. Effects of Lanthanide Shift Reagents on Some ^{13}C NMR Chemical Shifts

Compound	Shift Reagent	Shifts (ppm)[a] Induced at Carbon Atoms					
		1	2	3	4	5	6
Cyclohexanone	Eu(DPM)$_3$	48.9	17.2	11.5	7.5		
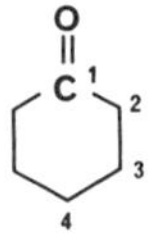	Pr(DPM)$_3$	−65.9	−27.9	−14.9	−10.5		
	Yb(DPM)$_3$	146.2	61.6	28.6	21.2		
Cyclohex-2-enone	Eu(DPM)$_3$	48.4	14.9	17.7	8.5	10.7	18.6
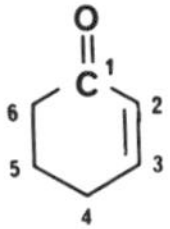	Pr(DPM)$_3$	−66.1	−26.2	−14.1	−8.5	−13.5	−28.9
4-t-Butylcyclohexanone	Eu(DPM)$_3$	47.9	16.0	9.5	5.3	2.5	0.9
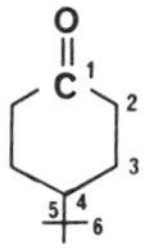	Pr(DPM)$_3$	−60.3	−23.9	−13.3	−8.7	−2.9	−1.8
	Yb(DPM)$_3$	149.7	61.7	28.4	20.7	9.0	5.5

Cyclohexanone (7)

Cyclohex-2-enone (8)

4-t-Butylcyclohexanone (9)

3-Methylbutan-2-one

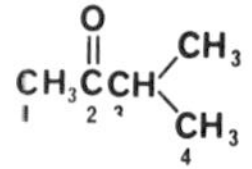

$CH_3CCH(CH_3)CH_3$ with O double bond; numbering: $CH_3(1)$ $C(2)$ $CH(3)$ $CH_3(4)$, $CH_3(1)$

(10)

Eu(DPM)$_3$	25.9	42.4	11.4	16.0
Pr(DPM)$_3$	−28.8	−58.3	−21.6	−18.8

2-Methylpentan-3-one

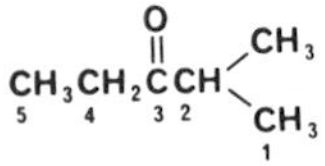

$CH_3CH_2CCH(CH_3)CH_3$ with O double bond; numbering: $CH_3(5)$ $CH_2(4)$ $C(3)$ $CH(2)$ $CH_3(1)$, $CH_3(1)$

(11)

Eu(DPM)$_3$	15.7	20.1	29.4	21.9	17.8
Pr(DPM)$_3$	−16.0	−22.1	−40.1	−22.1	−20.1

2-Hydroxy-3-methylbutane

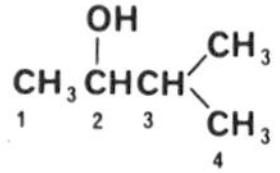

$CH_3CHCH(CH_3)CH_3$ with OH and CH_3; numbering: $CH_3(1)$ $CH(2)$ $CH(3)$ $CH_3(4)$

(12)

Eu(DPM)$_3$	23.5	60.1	15.3	14.1
Pr(DPM)$_3$	−42.3	−92.1	−36.0	−23.6

3-X-5-Methylhexane X = OH (13)   X = NH$_3$ (14)

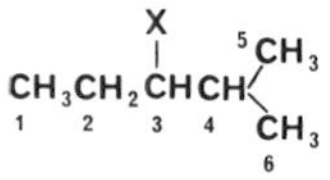

$CH_3CH_2CHCH_2CHCH_3$ with X and CH_3; numbering: $CH_3(1)$ $CH_2(2)$ $CH(3)$ $CH_2(4)$ $CH(5)$ $CH_3(6)$, $CH_3(5)$

X = OH
X = NH$_2$

Eu(DPM)$_3$	13.7	19.2	51.5	16.6	14.5	12.5
Pr(DPM)$_3$	−22.4	−35.1	−79.3	−34.6	−22.9	−21.4
Eu(DPM)$_3$	24.0	33.6	82.3	−11.8	17.6	18.3
Pr(DPM)$_3$	−31.1	−53.8	−114.1	−31.7	−28.5	−31.3

2-Amino-4-Methylbutane

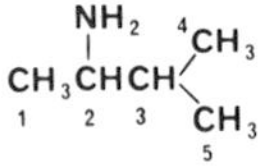

$CH_3CHCH_2CH(CH_3)CH_3$ with NH$_2$ and CH_3; numbering: $CH_3(1)$ $CH(2)$ $CH_2(3)$ $CH(4)$ $CH_3(5)$, $CH_3(4)$

(15)

Eu(DPM)$_3$	52.9	116.2	−30.7	18.8	19.6
Pr(DPM)$_3$	−72.1	−143.3	−31.5	−31.9	−28.4

Table 7.2. (Continued)

Compound		Shift Reagent	Shifts (ppm)[a] Induced at Carbon Atoms					
			1	2	3	4	5	6
1-X-Adamantane	X = OH (16)	Eu(DPM)₃	53.7	22.6	9.6	8.4		
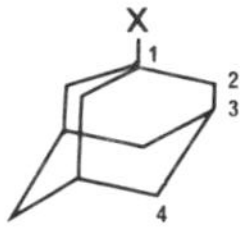		Pr(DPM)₃	-72.3	-32.6	-31.1	-11.1		
		Yb(DPM)₃	127.7	63.7	27.5	21.7		
	X = NH₂ (17)	Eu(DPM)₃	44.3	12.7	7.4	7.3		
		Pr(DPM)₃	-71.2	-28.9	-12.8	-10.3		
		Yb(DPM)₃	110.0	57.0	24.4	19.1		
Pyrrolidine (18)		Eu(DPM)₃	142.8	-8.2				
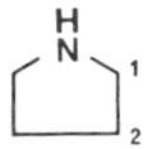		Pr(DPM)₃	-149.5	-40.0				

Source. Reference 102.

[a] Determined by sequential addition of shift reagent to the substrate solution and linear least-squares extrapolation to 1 : 1 (Ln) : (substrate) ratio; corrected, where necessary, for complex formation. A positive value corresponds to a paramagnetic (downfield) shift.

The addition of a paramagnetic relaxation reagent shortens considerably the experimental time required to obtain ^{13}C spectra of solutes at low concentrations or having long relaxation times, by allowing rapid pulse repetition rates. This is a tremendous benefit in the initial stages of spectrum measurement, particularly for finding quaternary carbon resonances, and in situations where great accuracy in chemical shifts is not required. Although the relaxation reagents are usually considered to be "shiftless," small changes in carbon chemical shifts, 0.1 ppm or less, have been reported[111] and observed in our laboratories. For accurate and precise chemical shift measurements, therefore, relaxation reagents may best be left on the laboratory shelf.

Isotopic Labeling

Deuteration

Spectral simplification made possible by substituting deuterium, ^{2}H, nuclei for protons is quite familiar to most chemists and NMR practitioners. In addition to removing some of the complexities of a spectrum by removing some of the resonances and their spin–spin couplings to other nuclei in the sample, deuterium substitution aids spectral assignments by affecting the chemical shifts and relaxation times of the observed nuclei, such as ^{13}C and ^{1}H. Deuterium NMR may also be observed directly; with broadband ^{1}H decoupling, simple spectra with ^{2}H chemical shifts that parallel the ^{1}H chemical shifts are obtained.

The benefits of deuterium substitution in assigning ^{13}C spectra of aromatic compounds are illustrated by its use in assigning 1- and 2-naphthyl compounds.[112] The proton-coupled ^{13}C spectra of some 1-substituted naphthalenes, **19**, and their 4-deutero analogs, and of 2-hydroxynaphthalene, **20**, and its

R = H, F, CN, CH$_3$
1-substituted-naphthalenes
(19)

2-hydroxynaphthalene
(20)

6-deutero analog were obtained. Deuterium isotope effects on ^{13}C chemical shifts of carbons up to six bonds away have been reported.[113] In the naphthyl compounds, the two-bond upfield deuterium isotope effect of approximately −0.1 ppm[114] allowed immediate identification of signals of carbons ortho to the deuterium—C-3 and C-10 in **19**, and C-1 and C-3 in **20**. One-bond deuterium isotope effects of about −0.3 ppm on the carbons bonded directly to the deuterium were also observed. Broadening caused by the three-bond ^{13}C–^{2}H spin–spin coupling, $^3J_{CD} \sim 1$ Hz,[115] allowed identification of the meta carbon resonances—C-2 and C-9 in **19** and C-8 and C-10 in **20**. Resonances caused by C-4 in **19** and C-6 in **20** were, of course, immediately assignable because of their characteristic 1:1:1 triplet structure, with $^1J_{CD} = 24$ Hz in **19**. Additionally, significant coupling between the deuterium at position 4 and C-5 in **19** allowed assignment of this carbon. Thus most of the carbon signals in these naphthalenes could be assigned unequivocally by deuterium substitution, making the remaining assignments relatively easy.

Similarly, assignment of the ^{13}C NMR spectrum of indazole, **21**, was made possible by deuterium labeling.[116]

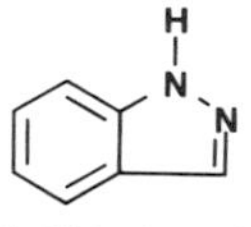

1-*H*-indazole
(**21**)

Labeling with the radioactive hydrogen isotope tritium, ^3H, is also possible. Tritium NMR chemical shifts, like deuterium chemical shifts, parallel those of protons. The ^3H spectra can thus sometimes aid in analysis of the corresponding ^1H spectra. For example, ^3H NMR was employed in a study of some labeled nitrogen heterocyclic compounds, to determine the regiospecificity and extent of labeling.[118] As an added benefit, the ^3H study enabled determination of the previously undetermined ^1H chemical shifts of phenanthridine, **22**.

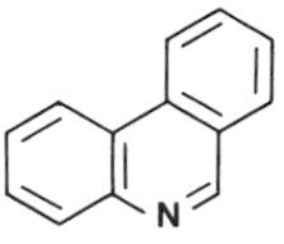

phenanthridine
(**22**)

Isotopic Enrichment

Considerable insight into the structure of molecules can often be obtained by enrichment of a naturally occurring magnetic isotope. A significant advantage of enrichment is the increased sensitivity that can be gained. For example, ^2H nuclei can be enriched by a factor of 6600 over their natural abundance (0.015%), or ^{13}C nuclei by a factor of 67, with consequent increases in sensitivity by the square of the enrichment factor.

Enrichment at specific sites may allow one to follow the detailed path of a single chemical moiety through a complex reaction. Carbon-13 enrichment has been used thus to establish biosynthetic pathways[119] and to investigate reaction mechanisms and reactive intermediates.[120]

If enrichment is uniformly distributed in the molecule and the percentage of enrichment is high, very complex homonuclear coupled spectra can be obtained. These spectra may yield useful information, however, in the form of spin-spin couplings, which afford insights into molecular structure and stereochemistry.

REFERENCES

1. T. C. Farrar and E. D. Becker, *Pulse and Fourier Transform NMR*, Academic, New York, 1971.

2. D. Shaw, *Fourier Transform NMR Spectroscopy*, Elsevier Scientific, Amsterdam, 1976.

3. E. D. Becker, *High Resolution NMR: Theory and Chemical Applications*, 2nd ed., Academic, New York, 1980.

4. E. Breitmaier and W. Voelter, *^{13}C NMR Spectroscopy*, 2nd ed., Verlag Chemie, Weinheim and New York, 1978.

5. J. B. Stothers, *Carbon-13 NMR Spectroscopy*, Academic, New York, 1972.

6. G. C. Levy and G. L. Nelson, *Carbon-13 Nuclear Magnetic Resonance for Organic Chemists*, Wiley-Interscience, New York, 1972.

7. L. M. Jackman and S. Sternhell, *Applications of Nuclear Magnetic Resonance Spectroscopy in Organic Chemistry*, 2nd ed., Pergamon, New York, 1969.

8. F. W. Wehrli and T. Wirthlin, *Interpretation of Carbon-13 NMR Spectra*, Heyden, London, 1976.

9. G. C. Levy and R. L. Lichter, *Nitrogen-15 Nuclear Magnetic Resonance Spectroscopy*, Wiley-Interscience, New York, 1979.

10. F. A. L. Anet, "^{13}C NMR at High Magnetic Fields" *in* G. C. Levy (Ed.): *Topics in Carbon-13 NMR Spectroscopy*, Vol. 1, pp. 209–227, Wiley-Interscience, New York, 1974.

11. F. A. L. Anet, D. L. Dalrymple, D. M. Grant, H. Hill, D. I. Hoult, L. F. Johnson, J. N. Shoolery, D. Terpstra, and A. P. Zens, "Experimental Techniques in ^{13}C NMR Spectroscopy" *in* G. C. Levy (Ed.): *Topics in Carbon-13 NMR Spectroscopy*, Vol. 3, pp. 2–16, Wiley-Interscience, New York, 1979.

12. W. Bruegel, *Handbook of NMR Spectral Parameters*, Heyden, London, 1979.

13. D. I. Hoult, *Prog. Nucl. Magn. Resonance* **6**, 51 (1977).

14. D. I. Hoult, *Prog. Nucl. Magn. Resonance* **7**, 145 (1978).

15. N. K. Wilson, "Nuclear Magnetic Resonance Spectroscopy" *in* K. G. Das (Ed.): *Pesticide Analysis*, pp. 263–328, Marcel Dekker, New York, 1981.

16. J. K. M. Sanders, *Annu. Rep. Prog. Chem. B* **75**, 3 (1978).

17. R. R. Ernst and W. A. Anderson, *Rev. Sci. Instrum.* **37**, 93 (1966).

18. R. R. Ernst, *J. Magn. Resonance* **3**, 10 (1970).

19. R. Kaiser, *J. Magn. Resonance* **3**, 28 (1970).

20. J. Dadok and R. F. Sprecher, *J. Magn. Resonance* **13**, 243 (1974).

21. R. K. Gupta, J. A. Ferretti, and E. D. Becker, *J. Magn. Resonance* **13**, 275 (1974).

22. J. W. Cooper, *in* P. R. Griffiths (Ed.): *Transform Techniques in Chemistry*, pp. 227–55, Plenum, New York, 1978.

23. R. D. Larsen, *in* P. R. Griffiths (Ed.): *Transform Techniques in Chemistry*, pp. 333–53, Plenum, New York, 1978.

24. T. C. Farrar, in P. R. Griffiths (Ed.): *Transform Techniques in Chemistry*, pp. 199–226, Plenum, New York, 1978.

25. E. D. Becker, J. A. Ferretti, and T. C. Farrar, *J. Am. Chem. Soc.* **91**, 7784 (1969).

26. R. L. Vold, J. S. Waugh, M. P. Klein, and D. E. Phelps, *J. Chem. Phys.* **48**, 3831 (1968).

27. N. K. Wilson, unpublished data.

28. R. Freeman and H. D. W. Hill, *J. Chem. Phys.* **54**, 3367 (1971).

29. J. Kowalewski, G. C. Levy, L. F. Johnson, and L. Palmer, *J. Magn. Resonance* **26**, 533 (1977).

30. R. K. Gupta, J. A. Ferretti, E. D. Becker, and G. H. Weiss, *J. Magn. Resonance* **38**, 447 (1980).

31. R. K. Gupta, *J. Magn. Resonance* **25**, 231 (1977).

32. J. C. Duplan, A. Briqnet, G. Tetu, and J. Delmau, *J. Magn. Resonance* **31**, 509 (1978).

33. H. Hanssum and H. Rüterjans, *J. Magn. Resonance* **39**, 65 (1980).

34. E. D. Becker, J. A. Ferretti, R. K. Gupta, and G. H. Weiss, *J. Magn. Resonance* **37**, 381 (1980).

35. J. L. Markley, W. J. Horsley, and M. P. Klein, *J. Chem. Phys.* **55**, 3604 (1971).

36. G. G. McDonald and J. S. Leigh, Jr., *J. Magn. Resonance* **9**, 358 (1973).

37. R. Freeman, H. D. W. Hill, and R. Kaptein, *J. Magn. Resonance* **7**, 82 (1972).

38. D. E. Jones, *J. Magn. Resonance* **6**, 191 (1972).

39. G. C. Levy and I. R. Peat, *J. Magn. Resonance* **18**, 500 (1975).

40. J. R. Lyerla, Jr. and D. M. Grant, *in* C. A. McDowell (Ed.): *International Reviews in Science, Physical Chemistry Series*, Vol. 4, Medical and Technical Publishers, London, 1972.

41. H. Y. Carr and E. M. Purcell, *Phys. Rev.* **94**, 630 (1954).

42. S. Meiboom and D. Gill, *Rev. Sci. Instrum.* **69**, 688 (1958).

43. I. Solomon, *Compt. Rend.* **248**, 92 (1959).

44. H. D. W. Hill and R. Freeman, *J. Chem. Phys.* **54**, 301 (1971).

45. G. Bodenhausen, R. Freeman, R. Niedermeyer, and D. L. Turner, *J. Magn. Resonance* **26**, 133 (1977).

46. S. J. Opella and J. S. Waugh, *J. Chem. Phys.* **66**, 4919 (1977).

47. E. F. Rybaczewski, B. L. Neff, J. S. Waugh, and J. S. Sherfinski, *J. Chem. Phys.* **67**, 1231 (1977).

48. D. Terpstra, "Two-Dimensional Fourier Transform ^{13}C NMR," *in* G. C. Levy (Ed.): *Topics in Carbon-13 Spectroscopy*, Vol. 3, pp. 62–78, Wiley-Interscience, New York, 1979.

49. R. Freeman and G. A. Morris, *J. Chem. Soc. Chem. Commun.* **1978**, 685.

50. R. Freeman and H. D. W. Hill, *J. Chem. Phys.* **54**, 301 (1971).

51. G. Bodenhausen, R. Freeman, and D. L. Turner, *J. Chem. Phys.* **65**, 839 (1976).

52. G. Bodenhausen, R. Freeman, R. Niedermeyer, and D. L. Turner, *J. Magn. Resonance* **24**, 291 (1976).

53. L. Müller, A. Kumar, and R. R. Ernst, *J. Magn. Resonance* **25**, 383 (1977).

54. R. Freeman, G. A. Morris, and D. L. Turner, *J. Magn. Resonance* **26**, 373 (1977).

55. G. Bodenhausen, R. Freeman, G. A. Morris, and D. L. Turner, *J. Magn. Resonance* **28**, 17 (1977).

56. G. Bodenhausen and R. Freeman, *J. Magn. Resonance* **28**, 303 (1977).

57. R. Niedermeyer and R. Freeman, *J. Magn. Resonance* **30**, 617 (1978).

58. J. Jeener, B. H. Meier, P. Bachmann, and R. R. Ernst, *J. Chem. Phys.* **71**, 4546 (1979).

59. G. A. Morris and R. Freeman, *J. Magn. Resonance* **29**, 433 (1978).

60. G. A. Morris and R. Freeman, *J. Am. Chem. Soc.* **100**, 6763 (1978).

61. H. Hill, "New Pulsed Excitation Methods," *in* G. C. Levy (Ed.): *Topics in Carbon-13 NMR Spectroscopy*, Vol. 3, pp. 84–101, Wiley-Interscience, New York, 1979.

62. R. Freeman, S. P. Kempsell, and M. H. Levitt, *J. Magn. Resonance* **38**, 453 (1980).

63. D. J. Burton, R. K. Harris, and L. H. Merwin, *J. Magn. Resonance* **39**, 159 (1980).

64. A. Pines, M. G. Gibby, and J. S. Waugh, *J. Chem. Phys.* **56**, 1776 (1972).

65. F. H. A. Rummens, "Proton Fourier Transform NMR as a Trace Analysis Tool: Optimization of Signal-to-Noise," *in* R. Haque and F. J. Biros (Eds.): *Mass Spectrometry and NMR Spectroscopy in Pesticide Chemistry*, pp. 219–237, Plenum, New York, 1974.

66. J. N. Shoolery, *Microsample Techniques in ¹H and ¹³C NMR Spectroscopy*, Varian Associates, Palo Alto, 1977.

67. M. Witanowski, L. Stefaniak, S. Szymanski, and H. Januszewski, *J. Magn. Resonance* **28**, 217 (1977).

68. R. L. Lichter, *J. Magn. Resonance* **18**, 367 (1975).

69. P. R. Srinivasan and R. L. Lichter, *J. Magn. Resonance* **28**, 227 (1977).

70. S. Brownstein and J. Bornais, *J. Magn. Resonance* **38**, 131 (1980).

71. A. L. VanGeet, *Anal. Chem.* **40**, 2227 (1968).

72. A. L. VanGeet, *Anal. Chem.* **42**, 679 (1970).

73. F. A. L. Anet, "Variable Temperature ¹³C NMR," *in* G. C. Levy (Ed.): *Topics in Carbon-13 NMR Spectroscopy*, Vol. 3, pp. 79–83, Wiley-Interscience, New York, 1979.

74. F. L. Dickert and S. W. Hellmann, *Anal. Chem.* **52**, 996 (1980).

75. J. A. Pople, W. G. Schneider, and H. J. Bernstein, *High-Resolution Nuclear Magnetic Resonance*, McGraw-Hill, New York, 1959.

76. R. J. Abraham, *Analysis of High-Resolution NMR Spectra*, Elsevier, Amsterdam, 1971.

77. A. A. Bothner-by and S. Castellano, LAOCN3, QCPE **10**, 111 (1967).

78. S. M. Castellano and A. A. Bothner-by, *J. Chem. Phys.* **41**, 3683 (1964).

79. J. A. Musso and A. Isaia, NMR-LAOCN-4A, QCPE **10**, 232 (1973).

80. F. J. Weigert and J. D. Roberts, *J. Am. Chem. Soc.* **89**, 2967 (1967).

81. V. G. Berezhnoi, A. N. Niyazov, and N. M. Sergeev, *Dokl. Akad. Nauk. SSSR* **228**, 1140 (1976).

82. N. K. Wilson and R. D. Zehr, *J. Org. Chem.* **43**, 1768 (1978).

83. N. K. Wilson and J. B. Stothers, *J. Magn. Resonance* **15**, 31 (1974).

84. R. R. Ernst, *J. Chem. Phys.* **45**, 3845 (1966).

85. F. J. Weigert, M. Jautelat, and J. D. Roberts, *Proc. Natl. Acad. Sci. U.S.A.* **60**, 1152 (1968).

86. E. R. Malinowski, *J. Am. Chem. Soc.* **83**, 4479 (1961).

87. R. Freeman and H. D. W. Hill, *J. Chem. Phys.* **54**, 3367 (1971).

88. B. Birdsall, N. J. M. Birdsall, and J. Feeney, *J. Chem. Soc. Chem. Commun.* **1972**, 316.

89. J. B. Grutzner and R. E. Santini, *J. Magn. Resonance* **19**, 173 (1975).

90. H. Günther, H. Schmickler, and G. Jikeli, *J. Magn. Resonance* **11**, 344 (1973).

91. H. Günther, H. Schmickler, H. Königshofen, K. Recker and E. Vogel, *Angew. Chem.* **85**, 261 (1973).

92. N. K. Wilson, *J. Am. Chem. Soc.* **97**, 3573 (1975).

93. J. H. Noggle and R. E. Schirmer, *The Nuclear Overhauser Effect*, Academic, New York, 1971.

94. H. Seto, T. Sasaki, H. Yonehara, and J. Uzawa, *Tetrahedron Lett.* **1978**, 923.

95. R. E. Sievers (Ed.): *NMR Shift Reagents*, Academic, New York, 1973.

96. K. A. Kime and R. E. Sievers, *Aldrichimica Acta* **10**, 54 (1977).

97. G. A. Webb, *Annu. Rep. NMR Spectrosc.* **6A**, 1 (1975).

98. A. F. Cockerill, G. L. O. Davis, R. C. Harden, and D. M. Rackham, *Chem. Rev.* **73**, 553 (1973).

99. J. Reuben, *Prog. Nucl. Magn. Resonance Spectrosc.* **9**, 1 (1973).

100. H. M. McConnell and R. E. Robertson, *J. Chem. Phys.* **29**, 1361 (1958).

101. B. L. Shapiro, J. R. Hlubucek, G. R. Sullivan, and L. F. Johnson, *J. Am. Chem. Soc.* **93**, 3281 (1971).

102. D. J. Chadwick and D. W. Williams, *J. Chem. Soc. Perkin Trans. II* **1974**, 1202.

103. E. Díaz, A. Guzmán, M. Cruz, J. Mares, D. J. Ramírez, and P. Joseph-Nathan, *Org. Magn. Resonance* **13**, 180 (1980).

104. O. A. Gansow, P. A. Loeffler, R. E. Davis, R. E. Lenkinski, and M. R. Willcott III, *J. Am. Chem. Soc.* **98**, 4250 (1976).

105. J. F. Desraux and C. N. Reilley, *J. Am. Chem. Soc.* **98**, 2105 (1976).

106. H. N. Cheng and H. S. Gutowsky, *J. Phys. Chem.* **84**, 1039 (1980).

107. J. A. G. Drake and D. W. Jones, *Spectrochim. Acta* **36A**, 23 (1980).

108. J. B. Stothers, C. T. Tan, and N. K. Wilson, *Org. Magn. Resonance* **9**, 408 (1977).

109. A. Dambska and A. Janowski, *Org. Magn. Resonance* **13**, 122 (1980).

110. R. Freeman, K. G. R. Pachler, and G. N. LaMar, *J. Chem. Phys.* **55**, 4586 (1971).

111. P. M. Henrichs and S. Gross, *J. Magn. Resonance* **17**, 399 (1975).

112. W. Kitching, M. Bullpitt, D. Doddrell, and W. Adcock, *Org. Magn. Resonance* **6**, 289 (1974).

113. F. W. Wehrli, D. Jeremic, M. Lj. Mihailovic, and S. Milosavljevic, *J. Chem. Soc. Chem. Commun.* **1978**, 302.

114. D. Lauer, E. L. Motell, D. Traficante, and G. E. Maciel, *J. Am. Chem. Soc.* **94**, 5335 (1972).

115. F. J. Weigert and J. D. Roberts, *J. Am. Chem. Soc.* **91**, 4940 (1969).

116. J. Elguero, A. Fruchier, and M. C. Pardo, *Canad. J. Chem.* **54**, 1329 (1976).

117. J. M. A. Al-Rawi, J. P. Bloxsidge, C. O'Brien, D. E. Caddy, J. A. Elvidge, J. R. Jones, and E. A. Evans, *J. Chem. Soc. Perkin Trans. II* **1974**, 1635.

118. J. A. Elvidge, J. R. Jones, R. B. Mane, and J. M. A. Al-Rawi, *J. Chem. Soc. Perkin Trans. II* **1979**, 386.

119. A. G. McInnes, J. A. Walter, J. L. C. Wright, and L. C. Vining, "^{13}C NMR Biosynthetic Studies" *in* G. C. Levy (Ed.): *Topics in Carbon-13 NMR Spectroscopy*, Vol. 2, pp. 125–178, Wiley-Interscience, New York, 1976.

120. J. B. Stothers, "^{13}C NMR Studies of Reaction Mechanisms and Reactive Intermediates" *in* G. C. Levy (Ed.): *Topics in Carbon-13 NMR Spectroscopy*, Vol. 1, pp. 229–286, Wiley-Interscience, New York, 1974.

Compound Index

Subject Index